Computational Intelligence Methods and Applications

Founding Editors

Sanghamitra Bandyopadhyay
Ujjwal Maulik
Patrick Siarry

Series Editor

Patrick Siarry, LiSSi, E.A. 3956, Université Paris-Est Créteil, Vitry-sur-Seine, France

The monographs and textbooks in this series explain methods developed in computational intelligence (including evolutionary computing, neural networks, and fuzzy systems), soft computing, statistics, and artificial intelligence, and their applications in domains such as heuristics and optimization; bioinformatics, computational biology, and biomedical engineering; image and signal processing, VLSI, and embedded system design; network design; process engineering; social networking; and data mining.

Kaushik Das Sharma • Subhajit Kar •
Madhubanti Maitra

Intelligent Computing in Carcinogenic Disease Detection

Springer

Kaushik Das Sharma
Department of Applied Physics, Electrical Engineering Section
University of Calcutta
Kolkata, West Bengal, India

Subhajit Kar
Department of Electrical Engineering
Institute of Engineering & Management
West Bengal, India

Madhubanti Maitra
Department of Electrical Engineering
Jadavpur University
Kolkata, West Bengal, India

ISSN 2510-1765 ISSN 2510-1773 (electronic)
Computational Intelligence Methods and Applications
ISBN 978-981-97-2426-0 ISBN 978-981-97-2424-6 (eBook)
https://doi.org/10.1007/978-981-97-2424-6

This Springer imprint is published by the registered company Springer Nature Singapore Pte Ltd.
The registered company address is: 152 Beach Road, #21-01/04 Gateway East, Singapore 189721, Singapore

To our families and beloved students.

Preface

This book analyzes some intelligent detection and classification methodologies for some specific carcinogenic diseases. The specific carcinogenic diseases include small round blue cell tumor (SRBCT), acute lymphoblastic leukemia (ALL), acute myeloid leukemia (AML), mixed-lineage leukemia (MLL), and lung cancer. The intelligent methodologies for diagnosis of these carcinogenic diseases have been developed on a strong foundation of gene-level, cell-level, and tissue-level data sets. Microarray gene expression profiles, microscopic blood smear images, and lung computed tomography images have been utilized in this purpose.

In recent years, microarray gene expression profiles play an important role in the diagnosis of different carcinogenic diseases. However, due to the availability of small number of samples compared to tens of thousands of genes, many existing computational methodologies failed to identify a small number of informative genes that would detect and classify different classes of carcinogenic diseases. Therefore, in this book, a particle swarm optimization (PSO)-based informative gene selection methodology with adaptive k-nearest neighbor (k-NN) classifier have been proposed for diagnostic prediction of different classes of carcinogenic diseases. A heuristic for selecting the appropriate number of neighbor (k) for k-NN classifier has also been proposed that would help to enhance the classification accuracy.

An intelligent methodology for detection and classification of ALL using microscopic blood smear image has also been proposed in this book. A large set of features have also been derived from the leukocytes for the intelligent diagnosis purpose. A weighted aggregation-based transposition PSO (WATPSO) has been proposed to select an optimized feature subset which is sufficient enough for classifying carcinogenic blast cells and healthy cells efficiently. The proposed WATPSO technique divides the main swarm into two sub-swarms, thereby allowing the sub-swarms to perform a heterogeneous search for achieving two different objectives. Here, finding a small subset of effective features and increasing the classification accuracy have been chosen as the two different objectives.

The value of the weights that are associated with these two objectives has been evaluated by introducing an adaptive weighted aggregation strategy. In the proposed weight selection strategy, the weights have been selected in such a way that classification accuracy increases with the help of a small discriminating feature subset. The proposed weight selection strategy has been embedded in group improvised harmony search (GrIHS) technique for intelligent detection and classification of lung nodules.

The methodologies, proposed in this book, can be used as a confirmatory option for the doctors for diagnostic prediction of carcinogenic diseases.

Kolkata, India
December, 2023

Kaushik Das Sharma
Subhajit Kar
Madhubanti Maitra

Contents

List of Abbreviations

ALL	Acute Lymphoblastic Leukemia
AML	Acute Myeloid Leukemia
ACS	American Cancer Society
ACO	Ant Colony Optimization
ANN	Artificial Neural Network
BF	Bacteria Foraging
BW	Band Width
cDNA	Complementary DNA
CT	Computed Tomography
CRT	Conformal Radiation Therapy
CAD	Computer-Aided Detection
CV	Cross-Validation
DNA	Deoxyribonucleic Acid
EWS	Ewing Sarcoma
EDTA	Ethylene Diamine Tetra-Acetic Acid
FNIH	Foundation for the National Institutes of Health
FDA	Food and Drug Administration
FN	False Negative
FP	False Positive
GLCM	Gray Level Co-occurrence Matrix
GrIHS	Group Improvised Harmony Search
GA	Genetic Algorithm
GMCR	Group Memory Considering Rate
GEO	Gene Expression Omnibus
HD	Hausdorff Dimension
HS	Harmony Search
HU	Hounsfield Unit
HM	Harmony Memory
HMS	Harmony Memory Size
HMCR	Harmony Memory Considering Rate
IMRT	Intensity-Modulated Radiation Therapy

IORT	Intraoperative Radiation Therapy
k-NN	*k*-Nearest Neighbor
LOOCV	Leave-One-Out-Cross-Validation
LIDC	Lung Image Database Consortium
LNCs	Lung Nodule Candidates
LBP	Local Binary Pattern
LDA	Linear Discriminant Analysis
mRNA	Messenger Ribonucleic Acid
MLL	Mixed-Lineage Leukemia
MRI	Magnetic Resonance Imaging
NHL	Non-Hodgkin Lymphoma
NB	Neuroblastoma
NCI	National Cancer Institute
NCC	Nearest Centroid Classifier
NNC	Nearest Neighbor Classifier
NBC	Naive-Bayes Classifier
PSO	Particle Swarm Optimization
PET	Positron Emission Tomography
PCA	Principal Component Analysis
PLSVIP	Partial Least Square Variant Importance in Projection
PLSIEG	Partial Least Square Independent-Variable Explanation Gain
PLNCs	Potential Lung Nodule Candidates
PAR	Pitch Adjustment Rate
RBC	Red Blood Cell
RFE	Recursive Feature Elimination
RMS	Rhabdomyosarcoma
SRBCT	Small Round Blue Cell Tumor
SVM	Support Vector Machine
SA	Simulated Annealing
SDM	Stimulating Discriminant Measure
SNO	Surface Normal Overlap
SNR	Signal-to-Noise Ratio
SE	Single Estimation
SFS	Successive Feature Selection
SI	Shape Index
TS	T-score
TP	True Positive
TN	True Negative
WBC	White Blood Cell
WATPSO	Weighted Aggregation-based Transposition Particle Swarm Optimization

Chapter 1
Introduction

Abstract This chapter provides an overview of carcinogenic diseases, commonly known as cancers. When cells in the human body undergo uncontrolled growth, they may form tumors, which can be either cancerous or benign. Carcinogenic cells, characterized by irregular shapes due to DNA mutations, have the ability to metastasis, or spread, to other areas of the body. Reducing mortality rates and providing effective treatment for cancer need early and precise cancer detection. While human perception was the basis for previous diagnosis approaches, new developments in digital imaging tools have brought in a new era where intelligent technology works in conjunction with the medical community. This chapter explores the methods used historically and intelligent aspects of detecting and classifying carcinogenic diseases, emphasizing the significance of early and accurate identification for improved patient outcomes.

1.1 Introduction

Carcinogenic diseases are generally considered to be the most dangerous diseases in the world. These carcinogenic diseases are basically known as Cancers, originated from the Greek word "karkinos" (used by a physician, Hippocrates, 460–370 B.C), generally used for describing the carcinogenic cells, and tissues or tumors [1]. The human body is composed of trillions of tiny cells. Normally, these cells expand and divide to multiply themselves to produce new cells. When the cells in a specific area of the human body start growing in an uncontrolled way, then they may form tumors that can be either cancerous or benign [2]. Sometimes, carcinogenic cells metastasize, wherein they spread to other portions of the human body through the bloodstream and/ or lymph and at the new sites these cells continue growing [3]. The carcinogenic cells are irregularly shaped due to carcinogenic mutations in DNA. The damage in the genetic information may sometimes be inherited, or in certain cases, may also be induced due to lifestyle factors [1]. Most cancers generally manifested themselves in the form of tumors. However, some other types of carcinogenic cells do not develop in the form of solid tumors.

K. Das Sharma et al., *Intelligent Computing in Carcinogenic Disease Detection*, Computational Intelligence Methods and Applications,
https://doi.org/10.1007/978-981-97-2424-6_1

With this brief introduction, even being non-medical persons, we can identify the alarming threats created by this deadly disease. Thus, for treating this life-threatening condition, it is a mandatory prerequisite that proper detection/diagnosis should be carried out at the earliest. However, the mortality rate of cancer patients can be reduced by a significant order of magnitude if the stage and state of the unfortunate malady can be identified without an error. Therefore, nowadays, we are facing a challenging situation that demands accurate and early detection of carcinogenic diseases followed by proper medical treatment as soon as possible.

A few decades past, analysis of diagnostic prediction of carcinogenic diseases has had rested on human perception. In recent years, because of the advances in digital imaging tools and techniques, the diagnostic science of carcinogenic diseases has entered into the most appealing phase where intelligent technology can effectively help human perception. Thus, gradually, the medical community has started to join hands together with the engineering fraternity to fight against this killer disease. In this context, this chapter highlights the historical approaches as well as the intelligent aspects of the detection and classification of carcinogenic diseases.

1.2 Historical Background of Carcinogenic Disease Detection

Since the primeval times, medical practitioners made various assumptions about the causes of occurrence of carcinogenic diseases and diagnosis of the same. In 1855, Virchow et al. suggested that tumors arise from embryo-like cells [4]. Later, Muller et al. experimentally demonstrated that carcinogenic tumors were formed by combining different other cells [5].

Subsequently, as a significant development, a nineteenth century medical researcher Rudolf Virchow framed and published specific autopsy protocols [6]. He also developed the concept of pathological processes. These works cited in the literature thus paved the path toward logical oncology, i.e., the study of carcinogenic diseases. Further, the improvements in anesthesia technology had supported the surgery procedure for treating carcinogenic diseases effectively [7].

In the nineteenth century, advanced magnifying instruments (microscopes) were introduced to concentrate on diseased tissues. In this context, Rudolf Virchow established the logical basis for the practice of the cutting-edge pathological investigation of infected cells using microscopes [4, 6, 7]. This advancement in magnification technology allowed a better realization of the cell intricacies and thus additionally helped the post-analysis phase of cancer surgery. The improved microscopic technology helped to analyze the body tissues, in detail, that had been operated by the surgeons, and the possibility of accurate diagnosis had further progressed.

A popular cancer screening method termed as "Pap Test" had been developed by Papanicolaou et al. in 1923 for early diagnosis of cervical cancer [8]. The American Cancer Society (ACS) has further encouraged the implementation of this screening technology globally in the 1960s. Modern mammography methodology had also been developed in the same period and had been recommended by the ACS in 1976 [1].

The art and science of cancer surgery have further been explored by Halsted et al. in the last decade of the nineteenth century with the invention of radical mastectomy [9]. The satisfactory local removal of malignant cells had been proposed by them as a way to treat carcinogenic diseases. Their proposal had been considered as the groundwork for cancer surgery for about a century [9]. Moreover, Paget et al. proposed that the carcinogenic cells that diffused in the bloodstream were capable to expand only in certain organs [10]. This proposal had turned into a key component in perceiving the restrictions of cancer surgery procedures. This proposal also enabled medical professionals to invent some important medications that would be capable of spreading all through the body to treat carcinogenic cells. During the twentieth century, modern surgical instruments, radiation, and chemotherapy had been invented to reduce the number of healthy cells that had been eliminated during the cancer operation [11]. In this context, some less invasive methodologies had also been proposed to destroy carcinogenic cells, for example, liquid nitrogen spray had generally been used in cryosurgery to kill abnormal cells [1]. Radio waves had also been proposed for the same purpose [12].

Further progress in cancer diagnosis had been made by the invention of medical imaging techniques such as ultrasound (sonography), computed tomography (CT scans), magnetic resonance imaging (MRI scans), and positron emission tomography (PET scans). CT images had been proposed to be used in conformal radiation therapy (CRT) for identifying the location of cancerous cells in three-dimensional space [13]. In CRT, the shape of the radiation beams was compared to the shape of the carcinogenic cells, and then it was delivered to the cells from different angles [13]. Intensity-modulated radiation therapy (IMRT) had further been introduced to control the intensity of the radiation beams [14]. This modified radiation methodology helped to fine-tune the dose of radiation on the carcinogenic cells. The radiation therapy had further progressed with the introduction of conformal proton beam radiation therapy. The proton beams had the advantage of causing minimum damage to the cells it passed through and maximum damage to the cells at the end of their path. This innovation had thus been very effective for applying high radiation to the carcinogenic cells while minimizing the damage to the surrounding cells. The type of radiation that was usually delivered at the time of surgery was termed as intraoperative radiation therapy (IORT) [14, 15]. During the time of surgery, the healthy cells could be shielded using the IORT so that the carcinogenic cells received the maximum dose of radiation [15].

The imaging technology was also being used in directing biopsy needles into the tumors. Sophisticated imaging instruments and miniature cameras had also been invented to guide the human body. For example, an endoscope had been utilized to treat tumors in the colon, throat, or bladder by entering the body through common openings like the mouth [12].

The brief history of the cause of carcinogenic diseases and related diagnosis tools is crucial in paving the way for optimal treatment plan selection. It might be inferred that cancer detection and classification necessitate a result with near

100% accuracy. With this in mind, the authors of this book have felt that along with conventional methodology, the advancement of machine learning and image processing technology could help in the diagnosis process by developing automatic detection mechanisms to support the manual diagnostic process and interpretation of histopathological reports.

Thus, this book aims to analyze different aspects of intelligent computing techniques in the diagnostic prediction of cancers and assess the performance of the same in terms of detection and classification accuracy.

1.3 Intelligent Computing Tools in Carcinogenic Disease Detection and Classification

Intelligent computing in carcinogenic disease detection and classification includes the application of different intelligent computing tools such as image processing tools, machine learning tools, and modern optimization tools on different biomedical data modalities. These different biomedical data modalities contain gene level data such as microarray gene expression data, cell level data such as microscopic blood smear images, and tissue level data such as x-ray images, ultrasound images, CT images, MRI images, etc. Hence, different image processing tools, machine learning tools, and, modern optimization tools have been introduced in this chapter to investigate their applicability to the diagnostic prediction of carcinogenic diseases.

1.3.1 Image Processing Tools

The preliminary step to analyze different biomedical image modalities is to process the images by several commonly used image processing tools such as thresholding, segmentation, region growing, etc. The thresholding technique has generally been used as an image pre-processing tool due to its simple mathematical concept. The background and foreground pixels of an image, I, are separated by using a simple threshold value, T. However, more than one threshold value, for example, T_1 and T_2, can be used to segment a particular pixel of interest $I(x, y)$ in one class if $T_1 \leq I(x, y) \leq T_2$, and in other class, otherwise [16]. The concept of more than one threshold can be modified further to multilevel thresholding by using multiple threshold values such as $T_1, T_2, \ldots, T_n$ for any integer value of n.

The pixels of an image can also be segmented using the region growing technique. Here, a particular pixel is generally selected as a seed point. Then the neighborhood pixels are compared to the seed point and added one by one to the seed point depending on their degree of similarity with the seed point. Thus, the region is keep growing with the addition of similar pixels one by one

[16]. The segmented image regions can further be processed using morphological operation. The morphological operation is useful for extracting various region components such as region shape, region boundary, etc. These operations are very useful for solving numerous image processing problems by using simple mathematical set theory concepts. This technique is also important to represent different characteristics of image regions such as region boundary, convex hull, signature, etc. [17].

However, image transformation techniques have been widely used in biomedical image processing applications. In this technique, a transformation function has generally been used to interpret the output image with respect to the input image. In this book, two popular image transformation techniques, distance transform and watershed transform, have been used to analyze the microscopic blood smear images. The distance transform technique is generally used in binary images where each pixel is replaced by the Euclidean distance of it to the nearest background pixel [18]. Hence, the output of a distance transformed image is a grayscale image where each pixel value represents the distance of that particular pixel from the boundary [16]. In contrast, watershed transform is generally applied to grayscale images. In this transform, the entire image is considered as a topographic surface. The intensity values represent the magnitude of elevations. Thus, the regions with similar intensity values represent the flat zone of the image. The lower intensity values are represented as regional minima. Two regional minimas are generally separated by a watershed line. The regions that are separated by the watershed line are also called catchment basins [16].

1.3.2 Machine Learning Tools

Generally, the machine learning tools require a set of input and output data. The tools learn from the input data to perform a specific task. The output data are predicted based on the learning analysis performed on the input data. Thus, the learning tools perform a specific task automatically and intelligently aiming to maximize the outcome efficiency of the learning process. However, the learning process can be categorized into supervised learning and unsupervised learning.

1.3.2.1 Supervised Learning and Unsupervised Learning

The supervised learning process uses both labeled input data and labeled output data. The labeled data samples are used to train the learning process. However, there exist some data samples that have never been used in the learning process, and these data samples are used as test data. In a nutshell, we can reiterate that the training data are used to train the learning algorithm, whereas the test data sets are used to evaluate the learning algorithm [19, 20]. Support vector machine (SVM), K-nearest neighbor (k-NN), etc., are examples of supervised learning algorithms.

In contrast, unsupervised learning algorithms do not use labeled data samples. In the learning process, similar input data samples clustered together. The unknown data samples are placed in the clusters depending upon their similarity measure with the clusters. For example, Euclidean distance may be used as a similarity measure for the data samples. The process is commonly known as clustering. C-means clustering and k-means clustering are popularly used as unsupervised learning algorithms [19, 20].

1.3.2.2 Cross-Validation and Blind Test

The process of learning during training can be improved by using the cross-validation technique. In this technique, the entire labeled training data samples are divided into several folds. For example, let the entire data samples be divided into K number of folds with approximately equal sizes of samples. At the time of the learning process, the Kth folder is used as test sample, and the remaining $(K - 1)$ folders are used as training samples. The learning process is then repeated K times with all the folds used as test samples separately. Thus, each input data sample is used as training as well as test samples in the learning process, and the average outcome is evaluated. As all the input data samples are used as training and test data samples repetitively, the process is termed as cross-validation. The process is termed as K-fold cross-validation if the number of folds is equal to K. The process is also termed as leave-one-out-cross-validation (LOOCV) if the number of folds is equal to the number of samples. In LOOCV, each input data sample is used as test data sample exactly once. In contrast, if some completely separated data samples, which have never been used to train the process, are used to evaluate the outcome of the cross-validated learning process, a blind test scenario is achieved.

1.3.2.3 Filter and Wrapper Methods

In machine learning, all the data samples possess some characteristics or attributes. These attributes are known as features. Generally, these features are used in the machine learning algorithms such as supervised and unsupervised learning. However, all the features are not equally important for use in machine learning algorithms. Some features may be redundant, and they do not contribute any additional information to the learning process. Thus, these redundant features increase the computational burden of the learning system. In this context, filter and wrapper methods have generally been used to eliminate redundant features [21, 22].

The filter methods are the feature ranking techniques where all the features are ranked according to their importance. The ranked features are then fed one by one into the machine learning algorithms to evaluate the outcome of the learning process [21]. In contrast to the feature ranking method, wrapper methods choose a certain set of features and evaluate their learning outcome. Thereafter, depending upon the merit of the outcome, it chooses a different set of features, and the process repeats

[22]. In this book, both the filter method and wrapper method have been thoroughly analyzed for intelligent detection of carcinogenic diseases.

1.3.3 *Modern Optimization Tools*

In this book, the use of modern optimization techniques as feature selection tools has also been analyzed. Modern optimization tools have generally been used to maximize or minimize a mathematical cost function in the presence of random variations of the input variables [23]. Some examples of these techniques include Genetic Algorithm (GA), Simulated Annealing (SA), Ant Colony Optimization (ACO), Particle Swarm Optimization (PSO), Harmony Search (HS), Bacteria Foraging (BF), etc. [23]. In this book, the optimization tools are used as a wrapper method for feature selection purposes. The tools are analyzed to find a small set of useful features that can significantly enhance the outcome of the learning process. In this book, different variants of modern optimization tools have also been analyzed in the context of an intelligent diagnosis of different carcinogenic diseases.

1.4 Organization of the Book

The book has been organized as follows.

Chapter 1 is the introductory concept, and Chap. 2 elucidates the biological backgrounds of the benchmark carcinogenic disease data sets. In Chap. 2, the gene level, cell level, and tissue level data sets have thoroughly been discussed. For the aim of intelligent detection, this chapter has also covered the biological importance and the preparation methods of these data sets.

Chapter 3 describes the applications of different intelligent computing approaches in the carcinogenic disease detection reported in the literature. The advantages and disadvantages of the different reported literature have also been discussed thoroughly in this chapter.

The classical approaches in gene evaluation for carcinogenic disease detection have been elaborated in Chap. 4. The filter method and wrapper method based on relevant gene identification techniques from microarray gene expression profiles have been discussed in this chapter. Some classical statistical approaches such as t-test, chi-squared test, and signal-to-noise ratio-based filter method and classical stochastic optimization such as PSO-based wrapper method have been analyzed for gene selection purposes.

The intelligent computing approach using PSO-based adaptive k-nearest neighborhood technique in microarray gene expression profile has been discussed in Chap. 5. The usefulness of PSO as a gene selection tool has been elaborated in this chapter. This chapter also analyzes the usefulness of an adaptive k-nearest neighbor classifier in enhancing the classification accuracy of the intelligent carcinogenic disease detection technique.

The intelligent computing approach using weighted aggregation-based transposition PSO (WATPSO) in cell level data has been discussed in Chap. 6. The detection and classification methodologies of leukocytes from microscopic blood smear images have been analyzed in this chapter. The WATPSO technique for the enhancement of leukocyte detection and classification performance has been elaborated thoroughly here.

The intelligent computing approach using group improvised harmony search-based adaptive weighted aggregation technique in tissue level data has been discussed in Chap. 7. In this chapter, the lung computed tomography images have been analyzed for the detection and classification of lung nodules. The usefulness of the group improvised harmony search technique in the context of feature selection has also been discussed here. The adaptive weight adjustment strategy has also been elaborated in the context of intelligent lung nodule detection and classification.

1.5 Summary

The chapter highlights that intelligent disease detection methodology is gradually entering the period of its advancement. In this context, it should be mentioned that biomedical research must strike a harmony between the well-sought automated methodologies for carcinogenic disease detection (inferred as intelligent methodologies) versus well-perceived human projections and interpretations. It has been assumed that the findings from this book will construct a solid path of intelligent carcinogenic disease detection and classification methodologies that could be accommodated by the doctors, radiologists, and hematologists as a solid second opinion to assist them in the accurate diagnosis of carcinogenic diseases.

References

1. Sudhakar A (2009) History of cancer, ancient and modern treatment methods. J Cancer Sci Ther 1:1
2. Morris H (1903) The Bradshaw Lecture ON CANCER AND ITS ORIGIN: Delivered at the Royal College of Surgeons on December 9th. Br Med J 2:1505
3. McDonald DM, Baluk P (2002) Significance of blood vessel leakiness in cancer. Cancer Res 2(18):5381–5385
4. Virchow R (1855) Editoral archiv für pathologische anatomie und physiologie und für klinische medizin, Vol. 167. Springer
5. Muller HJ (1938) Collecting Net 13:181
6. Virchow RLK (1896) Post-mortem examinations. Blakiston, Philadelphia
7. Harvey A (1976) Early contributions to the surgery of cancer: William S. Halsted, Hugh H. Young, and John G. Clark. Johns Hopkins Med J Supplement:69–83

8. Papanicolaou GN (1948) The cell smear method of diagnosing cancer. Am J Public Health Nations Health 38:202
9. Halsted W (1895) The results of operations for the cure of cancer of the breast performed at the Johns Hopkins Hospital from June 1889 to January 1894. Johns Hopkins Hosp Bull 4:297
10. Paget S (1889) The distribution of secondary growths in cancer of the breast. Lancet 133:571
11. Wyld L, Audisio RA, Poston GJ (2015) The evolution of cancer surgery and future perspectives. Nat Rev Clin Oncol 12:115
12. Huang X, Jain PK, El-Sayed IH, El-Sayed MA (2008) Plasmonic photothermal therapy (PPTT) using gold nanoparticles. Lasers Med Sci 23:217
13. Zelefsky M, Leibel S, Gaudin P, Kutcher G, Fleshner N, Venkatramen E, Reuter V, Fair W, Ling CC, Fuks Z (1998) Dose escalation with three-dimensional conformal radiation therapy affects the outcome in prostate cancer. Int J Radiat Oncol Biol Phys 41:491
14. Hall EJ (2006) Intensity-modulated radiation therapy, protons, and the risk of second cancers. Int J Radiat Oncol Biol Phys 65:1
15. Stathakis S (2010) The physics of radiation therapy. Med Phys 37:1374
16. Gonzalez R, Woods R (2002) Digital image processing, 2nd edn. Prentice Hall, Upper Saddle River, NJ
17. Soille P (2013) Morphological image analysis: principles and applications. Springer Science & Business Media, Berlin
18. Cuisenaire O (1999) Distance transformations: fast algorithms and applications to medical image processing
19. Kumar S (2004) Neural networks: a classroom approach. Tata McGraw-Hill Education, New York
20. Michalski RS, Carbonell JG, Mitchell TM (2013) Machine learning: an artificial intelligence approach. Springer Science & Business Media, Berlin
21. Lazar C, Taminau J, Meganck S, Steenhoff D, Coletta A, Molter C, de Schaetzen V, Duque R, Bersini H, Nowe A (2012) A survey on filter techniques for feature selection in gene expression microarray analysis. IEEE/ACM Trans Comput Biol Bioinf 9:1106
22. Xue B, Zhang M, Browne WN, Yao X (2015) A survey on evolutionary computation approaches to feature selection. IEEE Trans Evol Comput 20:606
23. Fouskakis D, Draper D (2002) Stochastic optimization: a review. Int Stat Rev 70, 315

Chapter 2
Biological Background of Benchmark Carcinogenic Data Sets

Abstract This chapter offers a thorough review of current research on the diagnostic prediction of carcinogenic diseases, with a focus on integrating patient-specific information at the gene, cell, and tissue levels. Advanced biological studies and microarray technology provide gene level data, which are essential for intelligent computing in the identification and classification of carcinogenic diseases. These data allow for the simultaneous monitoring of thousands of gene expression levels. Generally, cell level data, which are represented by microscopic blood smear images, have been visually analyzed by experienced technologists. An accurate diagnosis depends on identifying and detecting visual features that are impacted by the staining, smearing, and recording procedures. Tissue level data, exemplified by CT scan images, provide crucial insights into the distribution of physical attributes, essential for detecting and classifying carcinogenic diseases. This chapter explores the in-depth procedure for preparing CT scan, microscopic blood smear image, and microarray gene expression data. Additionally, standard benchmark data sets for these modalities are discussed, contributing to the advancement of research in the field.

2.1 Introduction

Present day research on diagnostic prediction of carcinogenic diseases largely depends on gene level, cell level, and tissue level information of patient body. Thus, in this chapter, a brief overview of gene level, cell level, and tissue level data has been provided.

Generally, the gene level data are derived from the biological experimental processes that can monitor the expression levels of thousands of genes at a particular condition simultaneously [1]. The simultaneous measurements of thousands of genes at a particular condition have been possible by the invention of microarray technology [2]. These microarray expression levels of thousands of genes play a crucial role in intelligent computing for carcinogenic disease detection and classification.

K. Das Sharma et al., *Intelligent Computing in Carcinogenic Disease Detection*, Computational Intelligence Methods and Applications,
https://doi.org/10.1007/978-981-97-2424-6_2

On the other hand, the examples of cell level data are the microscopic blood smear images. The analysis of these microscopic images is important for the diagnostic prediction of carcinogenic diseases [3]. Generally, the visual analysis has been performed by experienced laboratory technologists. The accurate diagnosis of carcinogenic diseases largely depends on the detection and identification of visual features of the blood cells. The characteristics of these visual features largely depend on the smearing, staining, and capturing process of blood smear images [4].

The examples of tissue level data are CT scan images. The CT images express the distribution of physical attributes in tissue level, whereas in x-ray images, the high-attenuating objects such as bones may hide other objects. Thus, the tissue level information obtained from CT scan images is important for detection and classification of carcinogenic diseases.

The preparation methodology of microarray gene expression data, microscopic blood smear image data, and CT scan data has been discussed in this chapter in detail. In this chapter, the standard benchmark microarray gene expression data, microscopic blood smear image data, and CT scan data have also been discussed.

2.2 Benchmark Carcinogenic Data Set I

In this book, the microarray gene expression data have been considered as carcinogenic data set I. The microarray technology is a very important breakthrough in experimental biology. With the aid of the DNA microarray technique, several gene expressions can be efficiently analyzed in a single experiment. Hence, the advancement in the microarray technology has allowed the researchers to measure the biological state of tens of thousands of genes at a particular time that is more particularly known as gene expression levels. Fundamentally, DNA microarrays are just tiny glass slides. DNA molecules are organized and adhered to the glass slides at spots or specific locations as shown in Fig. 2.1.

These spots are distributed throughout a microarray, and each spot includes a few million copies of the same DNA molecules that specifically correspond to a gene. The spots serve as probes in reality. The purpose of the probes is to measure the level of each gene's expression. The steps below are taken in order to measure the value of gene expression using a microarray [1]. First, molecules of messenger RNA (mRNA, the RNA molecules that transmit genetic information from the DNA to the ribosome) are gathered from a test sample (condition A) and a healthy sample (condition B) as shown in Fig. 2.1. Reverse transcriptase is subsequently utilized to transform the mRNA samples into complementary DNA (cDNA). Each cDNA sample is then labeled with fluorescent dyes of various colors. The two samples are then permitted to be combined. The process in which the cDNA molecules bind to the DNA probes on the slide is called hybridization. The laser is then used to activate the hybridized microarray and scan it at an appropriate wavelength to detect various dyes. For instance, the spot on the microarray will be red if the abundance of cDNA from condition A for a certain gene is higher than that from condition B. On the

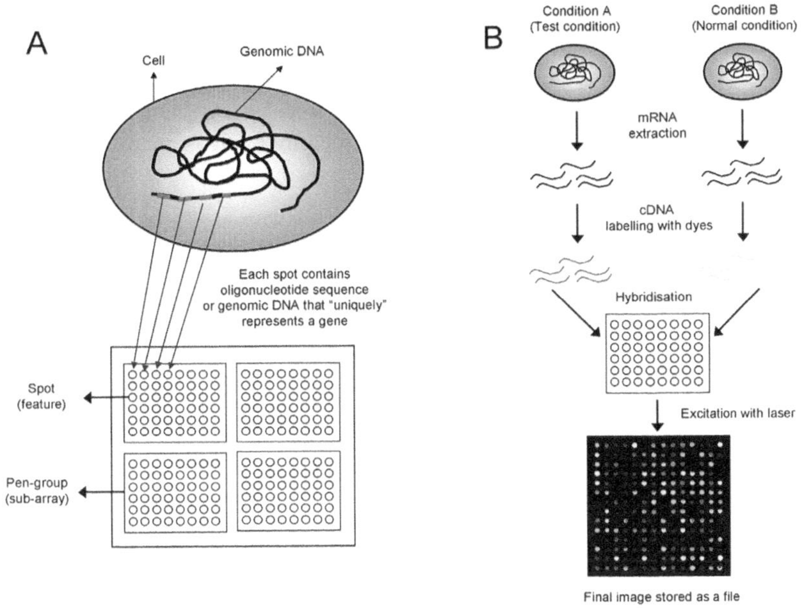

Fig. 2.1 Work flow of a typical expression microarray experiment. Figure courtesy: M. M. Babu, "Introduction to microarray data analysis," Computational genomics: Theory and application, vol. 17, no. 6, pp. 225–49, 2004

other hand, the spot in the microarray will be green. The spot will be yellow if the gene is expressed equally in both conditions, and black if the gene is not expressed equally in both conditions [1]. As a result, one eventually obtains an image of the microarray, in which each spot represents a gene and has an associated intensity value that represents the value of that gene's expression [5]. The work flow of a typical expression microarray experiment is shown in Fig. 2.1.

Therefore, in essence, the expression level of each gene is stored as an image. The employment of microarray scanners is the method used most frequently to analyze these images. To locate the spots, the images are examined. Actually, the microarray's dots are neatly organized into sub-arrays as shown in Fig. 2.1. Typically, microarray processing software is used to designate the location of each sub-array. The spot signal is then measured, and the background intensity is estimated, at a region inside the sub-array [6].

Generally, two methods are used to measure the spot signal. In the first technique, a fixed-size area is used in the spot's center of mass [7]. The alternative method, in contrast, simply takes into account the pixels within a spot's boundary when measuring the spot signal [7]. Though computationally expensive, the boundary pixel-based method typically results in a superior estimation of spot intensity. Then, to indicate spot intensity, the mean, median, and total of all pixel values are computed for both the spot and the background. Most microarray technologies typically calculate spot intensity using the median value of all the pixels. The amount of pixels to be included in the spot image for measurement is the most crucial element in the aforementioned process. Typically, the microarray scanner's default pixel size is 10 m. Nowadays, a lot of microarray scanners support 5 m pixels. The efficiency of the spot intensity measurement is increased by using smaller pixels [8].

2.2.1 Mathematical Representation of Microarray Gene Expression Data

The spot intensity measuring approach can be used to assess a gene's level of expression. The gene expression level is represented as a matrix. A specific gene is represented by each column, and a specific sample is represented by each row, in the matrix. The expression level for a gene for all the samples is cumulatively called as gene expression profile. Similar to this, a sample's overall gene expression levels are referred to as its sample expression profile [1].

Therefore, the expression profile for a particular gene is represented as a m-dimensional vector where m is the number of samples, and the expression profile of a sample is represented as a n-dimensional vector where n is the number of genes. Therefore, the gene expression matrix, X with n genes across m samples, is considered as a mXn matrix. $X = \begin{bmatrix} x_{11} & x_{12} & \dots & x_{1n} \\ x_{21} & x_{22} & \dots & x_{2n} \\ \dots & \dots & \dots & \dots \\ x_{m1} & x_{m2} & \dots & x_{mn} \end{bmatrix}$

The expression profile for gene i is represented as $G_i = \begin{bmatrix} x_{i1} \\ x_{i2} \\ \dots \\ x_{im} \end{bmatrix}$

The expression profile of sample j is represented as $G_j = \begin{bmatrix} x_{j1} & x_{j1} & \dots & x_{jn} \end{bmatrix}$

2.2.2 Benchmark Microarray Gene Expression Data Sets

SRBCT Data (http://research.nhgri.nih.gov/microarray/Supplement/)
The small round blue cell tumor data are commonly known as SRBCT data. It consists of four subclasses of cancers, namely, Ewing sarcoma (EWS), Non-Hodgkin lymphoma (NHL), Neuroblastoma (NB), and Rhabdomyosarcoma (RMS). SRBCT data contain 88 samples with 2308 number of genes in each sample. Out of 88 samples, 63 samples (EWS: 23, NHL: 8, NB: 12, RMS: 20) are used for training purpose, and the remaining 25 samples, which include 5 non-SRBCT samples [9]

and 20 SRBCT samples (EWS: 6, NHL: 3, NB: 6, RMS: 5), are used for the blind testing purpose of the intelligent detection algorithm [9].

ALL_AML Data (http://www.broadinstitute.org/cgibin/cancer/datasets.cgi)
This data set contains two subclasses of cancers, namely acute lymphoblastic leukemia (ALL) and acute myeloid leukemia (AML). The data set contains a total of 72 samples with 7129 number of genes in each sample. The training set containing 38 samples (ALL: 27, AML: 11) and the test set containing 34 samples (ALL: 20, AML: 14) are utilized by the intelligent detection algorithm [10].

MLL Data (http://www.biomedcentral.com/content/supplementary/1471-2105-7-228-S4.tgz)
This data set contains three subclasses of cancers, namely ALL, AML, and mixed lineage leukemia (MLL). There are a total of 72 samples each containing 12,582 numbers of genes. The training set contains 57 samples (ALL: 20, MLL: 17, AML: 20) and test set contains 15 samples (ALL: 4, MLL: 3, AML: 8) and are used by the intelligent detection algorithms [11].

2.3 Benchmark Carcinogenic Data Set II

In this book, the microscopic blood smear image data have been considered as carcinogenic data set II. Blood is a fluid that travels through the body with the aid of blood vessels. It helps to carry nutrients and oxygen to the tissues. There are several cells in it, including platelets, WBCs, and RBCs. The blood cell formation process is generally known as "hemopoiesis." The initial sources of blood cells are pluripotent stem cells, which underwent many phases of evolution before being constituted into specific cells of each kind and being able to enter the peripheral bloodstream (peripheral blood is the flowing or circulating blood of the body) [12].

In reality, blood smears are the thin films of peripheral blood. The creation of a peripheral blood smear is a laboratory process that entails the cytology (the study of a single cell type) of numerous blood cells spread out on a slide [13]. The peripheral blood smear is important in identifying a variety of blood-related disorders. The analysis of peripheral blood smears has been treated as a very important diagnostic tool for the hematologists. A high-quality smear and an accurate diagnostic interpretation of it could reveal important details about the clinical condition of the patients. This chapter has emphasized the fundamental science and technical skill involved in the peripheral blood smear.

Generally, visual microscopical examination provides a qualitative and quantitative analysis of blood smear [14]. Visual analysis of blood smear is an important tool for diagnosis and prevention of human blood-related diseases. Commercially accessible equipment have used flow cytometry techniques for automatic blood cell quantification since the 1980s (Beckman Coulter LH series, Sysmex XE-2100, Siemens ADV1A 120 and 2120) [13]. The method makes it possible to count blood cells and distinguish between various blood cell types. Additionally, the flow cytometry method calculates the blood's chemical and physical features, which aid in determining the type of cells [15].

To ensure effective results, pre-analytical parameters have to be controlled that can affect the characteristics of the film. These parameters include patient preparation, blood sampling technique, and sample preservation. The patient must receive the right counselling because the procedure is invasive. Typically, peripheral veins are used to draw blood first. After that, the blood is kept in an anticoagulant bottle. Blood to anticoagulant ratio needs to be maintained in the proper range. Generally, ethylene diamine tetra-acetic acid (EDTA) is used as an anticoagulant. When making the film, care should be made to prevent tissue injury to a minimum. The distribution of cells in the circulation can occasionally be affected by excess tissue fluid. It is advisable to evaluate the samples within two hours following sample collection. Degeneration of the blood's cellular components can result from a delay in blood smear preparation. This condition is frequently referred to as pseudo-thrombocytopenia (falsely reduced platelet count), which has developed as a result of platelet buildup [13].

Medical laboratory technologists with the necessary training make slides to evaluate the blood samples. Peripheral blood smear preparation often involves using the wedge or cover-slip method [16, 17]. In this procedure, a blood sample is put on a glass slide made of corrosion-resistant material, preferably at one end. The blood sample is then distributed onto the base slide using a spreader slide with chipped edges. The base slide and the spreader slide should be at an angle of 30 to 45°. The smear should have an oval, feathered end and should encompass two-thirds of the base slide length. The smear is then allowed to air dry before being stained to emphasize the morphological features of the cells. The process of blood smear preparation has been shown in Fig. 2.2.

The staining of blood cell is necessary as it is difficult to anticipate the cell structure due to inadequate contrast between nucleus and cytoplasm. By altering the color of the cell structures being studied, this cell staining approach increases

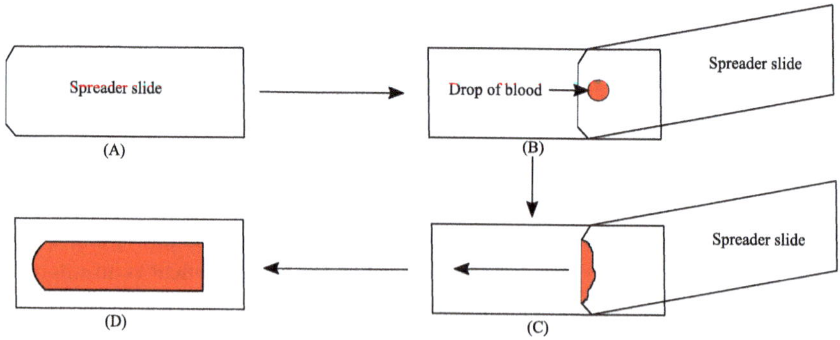

Fig. 2.2 Process of peripheral blood smear preparation. (**a**) Spreader slide. (**b**) Blood sample in glass slide. (**c**) Spreader slide is used to prepare blood smear. (**d**) Blood smear

contrast and makes it possible to see clearly. In the images of the microscopic blood smear, several stains can be applied [18]. Typically, blood sample analysis on slides involves hematoxylin and eosin staining. Hematoxylin and eosin, two stains, render the cytoplasm pink or red and the nuclei blue, respectively, to provide the contrast.

Blood microscopy requires a professional, methodical approach. A good smear may usually be examined in a few minutes. For a wider perspective and a different cell count, an abnormal film necessitates a longer exposure time. Blood cell counts can be performed using the peripheral blood smear [19]. In essence, it is impossible to overstate the significance of peripheral blood smear analysis in morphology and differential blood cell counting. It is important to pay close attention to the major blood cells' size, shape, color, and morphology. In shattered cell formations, there might be some anomalies, though. Blood films are typically analyzed alongside patient clinical information such as the patient's medical history and physical examination. The general components of the framework for interpreting peripheral blood smear images include results from other standard pathological tests, erythrocyte sedimentation rate, red cell indices, including whole blood cell count [13].

2.3.1 Microscopic Blood Smear Images

Labati et al. [20] have presented a public image database (ALL-IDB) of peripheral blood smear samples of healthy and cancerous patients that are specifically designed for the evolution of methodologies for segmentation and classification purposes [20]. A Canon PowerShot G5 camera and an optical laboratory microscope were used to take each image in the data set. All of the images have a 2592 × 1944 color resolution and are in the JPEG format. The images are captured with various microscope magnifications between 300 and 500. The ALL-IDB1 and ALL-IDB2 versions of the data set are each briefly discussed below.

The ALL-IDB1 data set consists of 108 images. It contains about 39000 elements of blood cell. The labeled blast cells present in the ALL-IDB1 data set are equal to 510. The ALL-IDB1 data set can be used for testing the segmentation capability and also for judging the classification accuracy of the proposed algorithm. The examples of the ALL-IDB1 data set are shown in Fig. 2.3 where the centroids of blast cells are marked as "×."

The ALL-IDB2 data set consists of the cropped area of interest of healthy and blast cells that originally belong to the ALL-IDB1 data set. This data set contains 260 images. Among them, 130 images belong to the healthy cells, and the remaining 130 images belong to the blast cells. The properties of the ALL-IDB2 data set are same as the ALL-IDB1 data set. The ALL-IDB2 data set can be used for testing the performance of the classification process. The representatives of the ALL-IDB2 data set are shown in Fig. 2.4.

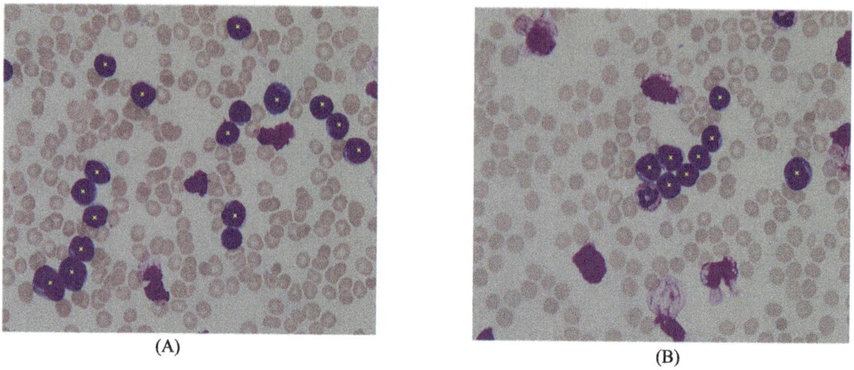

Fig. 2.3 (**a**) and (**b**) Some examples of ALL-IDB1 data set

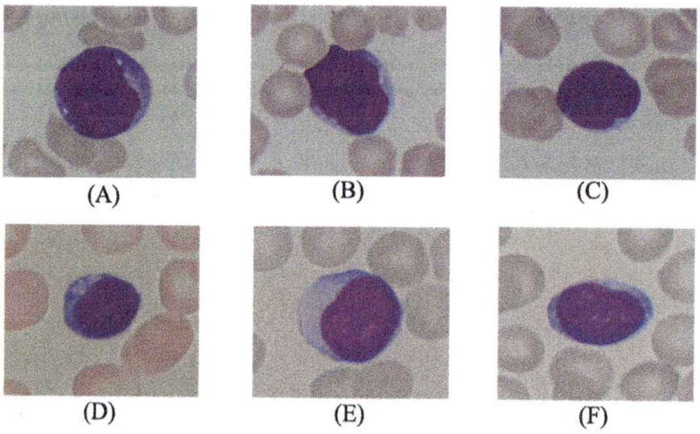

Fig. 2.4 Some examples of ALL-IDB2 data set. (**a**)–(**c**) Healthy cells, (**d**)–(**f**) annotated blast cells

2.4 Benchmark Carcinogenic Data Set III

In medical imaging, an image acquisition system has been used to record the distributions of physiological structures such as bones, tissues, cells, etc., of the human body. In modern days, the images are generally digitally captured. A computer-aided system must post-process these images in order to provide a more accurate analysis. In general, there are two types of medical images. Projecting a biological structure, such as a bone, onto a 2D image creates projection images. An example of a projection image is X-ray imaging. The physiological structures are superimposed on top of one another in this imaging approach [21]. As a result, the high-attenuating structures, such as bones, suppress the low-attenuating structures, such as soft tissues.

In contrast, computed tomography (CT) is a type of sliced imaging that takes pictures of the body in a number of different slices. Thus, the stack of numerous slices provides a 3D distribution of the human body's per volume unit [21, 22]. The CT technology was invented by G. N. Hounsfield in 1970s for capturing tissue

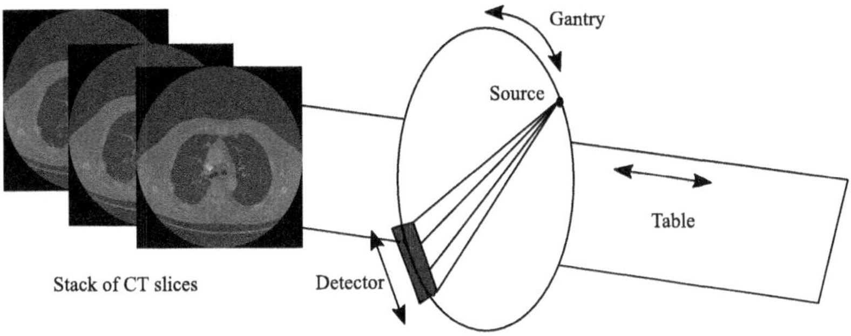

Fig. 2.5 Schematic view of CT image acquisition

level images [23]. A narrow beam of x-rays is focused at the patient during a computerized x-ray imaging procedure called a CT scan and is also spun around the patient's body. The method produces signals, which the computer then processes to produce image slices of the patient's body.

The CT scanner typically uses a motorized x-ray source that revolves around a circular framework known as the gantry. The patient is placed on a table that gently glides through the gantry as the CT scan is being performed. At the same time, the x-ray source rotates around the patient and directs the narrow beam of x-rays through the patient body as shown in Fig. 2.5. Additionally, CT scanners that are placed directly across from the x-ray source employ specialized x-ray detectors. The detector's job is to find and send the x-ray to the computer. The computer constructs a 2D image slice of the patient's body using complex mathematical methods following each rotation of the x-ray source [24]. The motorized table is advanced into the gantry after creating a 2D picture slice, and the x-ray scanning procedure is then repeated to create another 2D image slice. In this method, a patient's 3D image is created by stacking a number of 2D image slices in succession. The schematic view of the CT image acquisition process is shown in Fig. 2.5.

2.4.1 CT Images

In this book, the Lung Image Database Consortium (LIDC) [25] is used for the development of computer-aided diagnostic method for lung nodule detection and classification. This is a publicly available database and duly referenced by the medical imaging research community. The database is supported by National Cancer Institute (NCI), United States Department of Health and Human Services, and it is further advanced by the Foundation for the National Institutes of Health (FNIH), USA, accompanied by the Food and Drug Administration (FDA), United States Department of Health and Human Services.

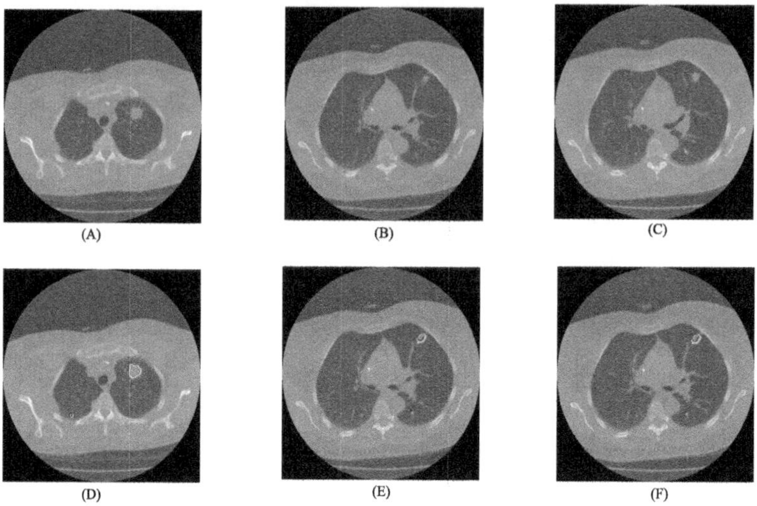

Fig. 2.6 Examples of lung nodules of LIDC database. (**a**)–(**c**) Lung nodules, (**d**)–(**f**) nodule outline by expert radiologists

1018 helical thoracic CT images from the seven participating institutions' picture archiving and communication systems (PACS) are available in the LIDC database. A total of 1010 distinct patients provided the 1018 scans. The nodule sizes in LIDC are restricted to roughly larger than or equal to 3 mm and less than 30 mm. For the image acquisition, many scanners are employed. For instance, 670 scans from the GE Medical Systems Lightspeed Scanner model, 74 scans from the Philips Brilliance scanner model, 69 scans from the Toshiba Aquilion scanner model, and 205 scans from the Siemens scanner model have all been collected. The slice thickness is between 0.6 mm and 5.0 mm. The reconstruction interval ranged from 0.45 to 5.0 mm [25]. Four radiologists independently review each scan in the database using a computer interface, and nodules are marked by them. Three types of nodule have been identified by them such as "nodule $\geq$ 3 mm," "nodule $<$ 3 mm," and "non-nodule $\geq$ 3 mm." The "nodule $\geq$ 3 mm" category is the database's main focus. The outer lines around the nodules in this group are thus drawn by each radiologist for each CT image. The examples of these nodules are shown in Fig. 2.6. Therefore, the database contains 1018 scans along with the same number of XML files containing the nodule outer line information.

2.5 Summary

The fundamental knowledge concerning data at the gene, cell, and tissue levels is provided in this chapter. This chapter has in-depth discussions on peripheral blood smear images, CT images, and microarray gene expression profiles as gene, cell, and tissue level data. These data collected from the databases can be very useful in the diagnosis and prognosis of cancer.

First, the microarray gene expression data have been studied in this book to identify the subset of useful genes that are primarily in charge of classifying the various types of malignancies. The subset of informative features responsible for the diagnosis and categorization of leukemia and lung cancer has then been identified through analysis of microscopic blood smear images and lung CT images. These features of intelligent cancer detection and classification approaches are highlighted in the next chapters.

References

1. Mm B (2004) An introduction to microarray data analysis. In: Computational genomics: theory and application. Horizon Bioscience, Cambridge, pp 225–49
2. Sotiriou C, Piccart MJ (2007) Taking gene-expression profiling to the clinic: when will molecular signatures become relevant to patient care? Nat Rev Cancer 7:545
3. Jenkins C, Hewamana S (2008) The blood film as a diagnostic too. Br J Hosp Med (2005) 69:M144
4. Abbas AK, Lichtman AH, Pillai S (2014) Cellular and molecular immunology E-book. Elsevier Health Sciences, Edinburgh
5. Yang YH, Dudoit S, Luu P, Lin DM, Peng V, Ngai J, Speed TP (2002) Normalization for cDNA microarray data: a robust composite method addressing single and multiple slide systematic variation. Nucleic Acids Res 30:e15
6. Bajcsy P (2004) Gridline: automatic grid alignment DNA microarray scans. IEEE Trans Image Process 13:15
7. Smyth GK, Speed T (2003) Normalization of cDNA microarray data. Methods 31:265
8. Angulo J, Serra J (2003) Automatic analysis of DNA microarray images using mathematical morphology. Bioinformatics 19:553
9. Shi Y, Eberhart R (1998) A modified particle swarm optimizer. In: IEEE International Conference on Evolutionary Computation Proceedings. IEEE World Congress on Computational Intelligence (Cat. No. 98TH8360). IEEE, Piscataway, pp 69–73
10. Mohapatra S, Patra D, Satpathy S (2014) Neural Comput Appl 24:1887
11. Kanazawa K, Kawata Y, Niki N, Satoh H, Ohmatsu H, Kakinuma R, Kaneko M, Moriyama N, Eguchi K (1998) An ensemble classifier system for early diagnosis of acute lymphoblastic leukemia in blood microscopic images. Comput Med Imaging Graphics 22:157
12. Shi Y, Eberhart RC (1998) Parameter selection in particle swarm optimization. In: International Conference on Evolutionary Programming. Springer, Berlin, pp 591–600
13. Zhang M-L, Zhou Z-H (2007) ML-KNN: A lazy learning approach to multi-label learning. Pattern Recognit 40:2038
14. Chen D, Chan KC, Wu X (2008) Gene expression analyses using genetic algorithm based hybrid approaches. In: 2008 IEEE Congress on Evolutionary Computation (IEEE World Congress on Computational Intelligence). IEEE, Piscataway, pp 963–969
15. Vandeginste BM, Massart D, Buydens L, De Jong S, Lewi P, Verbeke J (1998) Handbook of chemometrics and qualimetrics. Data handling in science and technology, vol. 20B, Elsevier
16. Dudoit S, Fridlyand J, Speed TP (2002) Comparison of discrimination methods for the classification of tumors using gene expression data. J Am Stat Assoc 97:77
17. Tarlow D, Swersky K, Charlin L, Sutskever I, Zemel R (2013) Stochastic k-neighborhood selection for supervised and unsupervised learning. In: International Conference on Machine Learning. PMLR PMLR, pp 199–207

18. Ougiaroglou S, Nanopoulos A, Papadopoulos AN, Manolopoulos Y, Welzer-Druzovec T (2007) Adaptive k-nearest-neighbor classification using a dynamic number of nearest neighbors. In: East European Conference on Advances in Databases and Information Systems. Springer, Berlin, pp 66–82
19. Engelbrecht AP (2007) Computational swarm intelligence. In: Computational intelligence: an introduction. John Wiley & Sons, pp 285–412
20. Labati RD, Piuri V, Scotti F (2011) All-IDB: The acute lymphoblastic leukemia image database for image processing. In: 2011 18th IEEE International Conference on Image Processing. IEEE, Piscataway, pp 2045–2048
21. Toennies KD (2017) Guide to medical image analysis. Springer, Berlin
22. Henschke CI, McCauley DI, Yankelevitz DF, Naidich DP, McGuinness G, Miettinen OS, Libby DM, Pasmantier MW, Koizumi J, Altorki NK et al (1999) Early Lung Cancer Action Project: overall design and findings from baseline screening. Lancet 354:99
23. Buzug TM (2009) Computed tomography: from photon statistics to modern cone-beam CT. Soc Nuclear Med 50(7)
24. Geluk R (1979) Transverse Analogue Tomography (TAT). Opt Act Int J Opt 26:1367
25. Armato III SG, McLennan G, McNitt-Gray MF, Meyer CR, Yankelevitz D, Aberle DR, Henschke CI, Hoffman EA, Kazerooni EA, MacMahon H et al (2004) Lung image database consortium: developing a resource for the medical imaging research community. Radiology 232:739

Chapter 3
Intelligent Computing Approaches for Carcinogenic Disease Detection: A Review

Abstract This chapter provides an overview of recent advancements in the identification and classification of cancer-causing disorders, focusing on intelligent diagnosis through the application of diverse machine learning techniques. In the last few years, researchers have developed advanced methods that make use of gene, cell, and tissue databases. The chapter provides an overview of the different approaches used in this sector and provides insight into how intelligent technologies are changing the landscape of cancer detection. The significance of these strategies is analyzed, offering insights into how they might improve diagnosis accuracy and advance our understanding of cancer-related disorders.

3.1 Introduction

In the past several years, several sophisticated approaches for identifying and categorizing cancer-causing disorders have been put out in the literature. For the objective of intelligent diagnosis, the researchers have used various machine learning techniques on data sets at the gene, cell, and tissue levels. This chapter describes different strategies that have been published in recent years in this specific field of research and provides a brief explanation of their importance.

3.2 Intelligent Computing Approaches for Benchmark Carcinogenic Data Set I

In terms of microarray innovation, the diagnostic prediction of various cancer types has received a lot of attention recently. The advancement of the microarray can overcome the limitations of traditional methods. Traditional methods of cancer diagnosis mostly rely on the physical characteristics of malignancies. However, it is exceedingly challenging to find a definite distinction between various cancer kinds

K. Das Sharma et al., *Intelligent Computing in Carcinogenic Disease Detection*, Computational Intelligence Methods and Applications,
https://doi.org/10.1007/978-981-97-2424-6_3

as suggested by their morphological manifestations. The researchers in this domain, therefore, choose the recently developed microarray technique [1].

The following characteristic of gene expression data makes it difficult to predict malignancies using gene expression-based diagnostics. First, gene expression data is typically high dimensional. There are typically many thousands to over ten thousand dimensions. Second, there are usually insufficiently large gene expression profile samples. Therefore, the minimal number of training samples and the high dimensional input gene expression profile present challenges for researchers. However, several redundant genes can be found in this huge gene pool. These redundant genes greatly worsen the accuracy of cancer classification. Hence, choosing a subset of informative genes is essential for enhancing the accuracy of cancer diagnosis and prediction.

Filter method-based approaches and wrapper method-based techniques are the two main strategies for gene selection. Gene selection is carried out using filter-based methods prior to classifier application. Typically, they employ gene ranking strategies based on statistical tools, with the better ranked genes serving as the input to a classification scheme. Wrapper-based techniques, in contrast, combine a categorization scheme with gene selection. Using a learning system, they look for a beneficial subset of genes [2].

In 1999, Golub et al. revealed that mathematical or statistical methods might be used to produce the diagnostic prediction of tumors [2]. According to the strength of the correlation, all the gene expression levels in their study had been ranked. They investigated the neighborhood analysis approach to quantify the association between gene expression and class prediction. They had also suggested a strategy that predicts a gene's expression levels using a list of the best-positioned instructive genes. Every gene expressed a vote for a particular class in their suggested approach. By comparing a gene's specific expression level to the average expression level of its class, the vote had been expressed. Each vote was then weighted by the positioning value and the class was appointed by the separate voting aggregates for different classes. To prevent any experimental artefacts, their methodology needed careful experimental design. This could lead to the identification of incorrect classifications.

An artificial neural network (ANN)-based method for categorizing tumors into specific diagnostic categories was proposed by Khan et al. in 2001 [3]. A process of error reduction through learning from experience had been employed to employ ANN as a pattern identification tool. Gene expression levels were used to modify the ANN's parameters, and the output was divided into various groups. The ANN model had determined the role played by each gene in the categorization system. Their suggested approach only used the ANN model's genes for which accurate evaluations could be made across all data. It is therefore very likely that some genes that are substantially expressed in some cancers would be overlooked.

To categorize tumor samples into two classes, in 2001, Hedenfalk et al. employed a class prediction approach [4]. To rank the genes according to their significance, they utilized the F-test, weighted gene analysis, and mutual information scoring. The relationships between statistically significant genes had been evaluated using gene ranking approaches. The association of gene expression levels in cancer subgroups had also been examined using multidimensional scaling. Only a few of the top-ranked genes were used in their suggested methodology to evaluate all samples. As a result, there is a strong likelihood that some genes that are highly expressed in particular cancer subgroups may be overlooked.

To address the class prediction challenge, Tibshirani et al. suggested a shrunken centroid-based gene selection mechanism in 2002 [5]. They used the distance between the centroid for each class and the centroid for all the gene expression data to obtain the shrunken centroid. The centroids that were closer to the global centroid than a set threshold have been eliminated. There was a correlation between the expression levels of several genes belonging to a given class in the gene expression profiles. It is necessary to search for helpful genes that have a strong correlation with one another. Moreover, various genes in the same biological pathway could interact with one another. To detect the interaction, a thorough analysis is required.

Guyon et al. introduced a recursive feature elimination (RFE)-based gene ranking approach in 2002 to rate the genes [6]. They recommended taking the least informative gene out of the training set and repeating the process until the gene ranked list was finished. Their experimental outcomes had been favorably compared to those in [2]. The RFE-based gene ranking approach that was suggested offers a basis for linear classifiers. The method can be improved by applying nonlinear classifiers, various kernel techniques, clustering strategies, etc.

Chu et al. suggested a gene selection technique utilizing Gaussian processes in 2005 [7]. Their approach has the benefit of an explicit probabilistic framework that allows for the evaluation of the degree of uncertainty in the gene expression profile. Also, a forward selection technique was used to identify a small group of useful genes. In reality, their concept focuses primarily on binary classes of gene expression data, which tends to identify core genes or discrete biomarkers.

Fu et al. introduced a probability analysis method in 2005 to assess the relevance of each gene by taking into account both statistical significance and performance concerns [8]. The repeatability of the chosen genes had been evaluated using reliability analysis in conjunction with gene ranking. Their strategy had also been contrasted with other well-liked approaches put out by Khan et al. and Tibshirani et al., and they discovered strong agreement between the outcomes [3, 5]. They suggested that reliability analysis should be based on the effectiveness of gene selection, and that the approach might be expanded to enhance methods of performance-based gene selection.

The RFE-based gene selection method was modified by Duan et al. in 2005 [9]. Their suggested strategy used statistical analysis to calculate the gene ranking score at each stage of the RFE algorithm. The weight vectors of several linear SVMs were the foundation for their statistical analysis. The suggested algorithm's training phase had utilized the replication of samples in a number of linear SVMs. Also, they used the approach suggested by Golub et al. to determine the sensitivity of genes [2]. The use of several SVM classifiers made their proposed method computationally expensive.

Yang et al. used the replication technique of samples for a stable gene selection in 2006 [10]. They had combined the replication method with a gene score system that takes into account both intra- and inter-class variation as well as the mean across different classes. As no particular statistical model was necessary, their suggested approach was model-free. Also, they utilized a *k*-NN classifier for classification purposes, with *k* set to 5. The reason behind the selection of this particular value of *k* for *k*-NN classifier had not been addressed by them.

Kestler et al. utilized the same technique as Tibshirani et al. in 2006 and identified some misclassification by using a limited number of genes [5, 11]. The classification error rate had been assessed using the genes chosen by their technique.

In 2007, Tang et al. used a two-stage SVM-RFE algorithm to improve the methods proposed by Guyon et al. [6, 12]. It had been suggested to maintain the important genes and remove the extraneous ones in stages. The process was repeated numerous times to obtain different gene subsets, which were then uniquely combined to form a gene subset. Another SVM-RFE method with the stringent filter-out factor was used in the second stage. The performance of their suggested work was superior to the earlier reported works by Golub et al. and Hedenfalk et al [2, 4]. The fact that only one redundant gene had been removed at a time at each level of the algorithm made their suggested solution computationally expensive as well.

The aforementioned methods have, however, assessed the significance of specific genes in relation to the target class. The interaction between various genes, which may be helpful for the categorization task, has been disregarded [13]. Moreover, research has revealed another problem with gene ranking strategies. The best subset of genes in terms of classification performance is not produced by combining the top-ranked genes. When combined with other genes, the least-ranked genes may also be beneficial. Several approaches have been developed in the literature to improve the performance of the diagnostic prediction task in order to address the aforementioned disadvantages [14–17].

Moreover, the filter methods do not account for feedback on classification performance at the time of gene ranking. This has been identified as a significant flaw in filter-based gene selection approaches. Wrapper method-based gene selection techniques are typically employed to address this disadvantage. The wrapper approaches use classification performance feedback to accomplish gene selection. A classifier is trained with training data, and its performance is evaluated with test samples to determine the classification performance for one subset of the selected genes. Depending on the effectiveness of the classification, the process is repeated iteratively. Moreover, the computational complexity of wrapper method-based gene selection has been challenged. Due to this reason, the filter methods have been preferred over wrapper methods while the wrapper methods usually produce better classification performance.

Inza et al. in 2002 [17] employed three traditional wrapper approaches: sequential forward and backward selection [14], floating selection [15], and best-first search [16]. They used the Naive Bayes rule and 1-nearest neighbor as a classifier to get acceptable results on microarray data sets. The quality of each gene subset had been assessed through the use of traditional wrapping methodologies. The biological significance of the chosen genes had not been attempted to be described in this work. Moreover, their suggested strategy yielded a unique gene subset. Hence, a collection of various gene subsets is required in order to create a coherent gene subset. In this context, the classical wrapper approaches have been modified by the use of stochastic search algorithms.

A GA-based technique based on the Markov blanket concept was proposed by Zhu et al. in 2007 [18, 19]. To fine-tune the gene selection process, the Markov blanket idea added or eliminated genes from the GA solution. Their suggested approach eliminated the redundant genes while keeping the pertinent ones, utilizing both the Markov model and GA to increase classification accuracy. The GA-based gene selection idea used in their proposed method places restrictions on the size of the informative gene subset in each chromosome. This limits the space of all the genes on the microarray where an optimal informative gene subset can be found.

Li et al. in 2007 presented a tabu search-based wrapper approach for high dimensional microarray gene selection [20]. The tabu search method had made use of memory to direct the search in the direction of the best outcome. Each iteration of tabu search had resulted in the optimal solution being shifted closer to the neighborhood. The movements were kept in memory. In the tabu search-based gene selection method, neighborhood size is a crucial consideration. The comparison between this study and the GA-based approach demonstrates that tabu search identifies fewer genes than GA. Nevertheless, because tabu search only applies to one gene at a time, it is computationally expensive to locate the best gene subset. Moreover, their suggested strategy does not account for gene interactions.

Shen et al. presented the hybrid PSO-tabu search methodology in 2008 to choose the most informative genes from a subset of those expressed in microarray gene expression data [21]. The tabu search method used a memory system to have the capacity to avoid local minima. Although the tabu search method's convergence speed was slow, the addition of PSO enabled it to locate an optimal solution with a very short computing time. They also used a filter approach called the t-test [22] to rank the genes initially according to their significance. The top genes determined by the t-test approach have subsequently been applied to the proposed hybrid PSO-tabu search methodology. As a result, the PSO-tabu search engine excluded a significant number of genes that were not included in the top rank list, and their proposed wrapper approach was limited to only a few top-ranked genes.

In 2008, Li et al. also used a hybrid strategy that included PSO and GA [23]. To improve the performance of gene selection, the GA parameters had been added to the PSO. Prior to the implementation of the hybrid PSO-GA technique, they used the Wilcoxon rank sum test. As a result, this hybrid technique was restricted to a small number of top-ranked genes determined by the Wilcoxon rank sum test.

In 2011, Mohamad et al. proposed an improved binary PSO-based gene selection algorithm [24]. They had introduced the idea of particle speed to obtain a near optimal solution. Prior to using their suggested wrapper methods, they also used a filter mechanism called gain ratio [24]. They had evaluated LOOCV accuracy to measure the performance of their proposed work. Moreover, only the top 500 genes could be used with the suggested strategy, and no blind test accuracy had been reported.

In 2013, Li et al. modified the hybrid PSO-GA strategy by employing GA search space slicing PSO to locate the global optima in the subspaces [25]. They developed the search space modification method to help the proposed algorithm's global convergence. The search space had been divided up into several smaller subspaces in their proposed method. Then the fitness of best particle of each subspace had been used by GA to direct the PSO to search for promising subspaces. In this way, they had utilized the faster convergence property of PSO while keeping the global search capacity of GA. In order to slice the search space and separately search each subspace, a large computing time is necessary. Thus, their proposed method suffers from computational overhead.

A PSO-based gene selection technique and a C4.5 decision tree classifier were further merged by Chen et al. in 2013 [26]. In their suggested method, they had also calculated the accuracy of fivefold cross-validation. The classification accuracy attained using their suggested methodology, however, was not very high, and they had not provided any information regarding the accuracy of blind tests.

Several heuristics-based wrapper approaches have been used in the reported literature [18, 20, 21, 23–26] to treat the reduced number of genes that have been produced as a result of filter methods. In reality, this means that a small number of genes have been regarded to play a significant role in the wrapper-based gene selection approaches whereas a vast number of genes have been completely ignored. As a result, the gaps in the prior techniques are noted as follows. The heuristic-based wrapper strategy requires a large number of genes to participate, and blind testing on the data set is also required to assess how well the identified genes performed in classification. Further, the usefulness of the selected genes is supposed to be reconfirmed by using a different classifier.

3.3 Intelligent Computing Approaches for Benchmark Carcinogenic Data Set II

Cell level information is also crucial for cancer diagnosis and prediction, in addition to microarray gene expression profile-based methods. In order to identify various blood cell types, microscopic blood smear images can be a crucial source of cell level information. White blood cells, also known as leukocytes, play a major role in the diagnosis of leukemia. Nowadays, the analysis of leukocytes is carried out by expert hematologists by examining the cell level microscopic blood smear images [27]. However, the hematologist's skill and competence have a significant role in how precisely leukocytes are found and classified using this manual procedure. Hence, an intelligent leukocyte detection and categorization system is truly needed as an additional predictive mechanism to aid hematologists in their diagnostic prediction.

In recent years, many research works have been devoted to the detection and classification of leukocytes [27–36]. An automatic method for counting blood cells was presented by Sinha et al. in 2003 [27]. They have employed the k-means clustering technique to separate the leukocytes from the background image. For the classification of leukocytes, they have also retrieved several attributes based on shape, color, and texture. However, their automatic methodology suffers from low segmentation accuracy as well as low classification accuracy.

In 2004, classification of subtypes of leukocytes was proposed by Piuri et al. [28]. Several traditional image processing techniques, such as contrast stretching, opening, edge detection, hole filling, etc., were used to segment the leukocytes. They also studied many morphological characteristics in order to classify them. They developed a method to count the lobes on the nucleus to distinguish between various leukocyte subtypes. They do not report very high classification accuracy. Additionally, a completely automated system requires to analyze tumor deformation in cell morphology.

Leukocyte segmentation and background separation were also put out by Scotti et al. in 2006 [29]. They employed histogram thresholding and fuzzy k-means clustering for the segmentation process. The authors also suggested a self-adaptive segmentation method to recognize the leukocytes [29]. This technique made it simple to extract characteristics from leukocytes. The segmentation of fragmented leukocytes, which frequently lack a nucleus, is a drawback of their suggested procedure.

Rezatofighi et al. proposed the use of the local binary pattern (LBP) in 2011 to distinguish between blast leukocytes and healthy leukocytes [30]. Two characteristics that represent the texture's structure and changes in grey levels were extracted using LBP. Prior to calculating the LBP feature, they had proposed a segmentation technique that is based on the Gram-Schmidt orthogonalization method [30]. The effectiveness of two classifiers, such as ANN and SVM, had also been compared. The proposed approach is resistant to changes in color and brightness of the input images. Additionally, since blast cells and other cells can appear in peripheral images, classification accuracy needs to be ımproved.

In 2012, Madhloom et al. proposed a color-based nucleus segmentation technique for peripheral blood images [31]. They used a feature extraction method based on principal component analysis (PCA) to identify various leukocyte types. The proposed methodology's disadvantage is that only features that have been retrieved from the cell nucleus are taken into account. In addition, the cytoplasm of the cell and the entire leukocyte can be used to extract the features, which can then be used to identify the leukocytes.

Mohapatra et al. suggested an independent method for segmenting and identifying lymphocytes using shadowed C-means clustering in 2013 [32]. The suggested method has also been used to segment the cytoplasm and nucleus of cells independently. Then, they retrieved the features that were based on shape and texture. They used t-test, a statistical feature selection tool, to look for valuable features. Both benign and malignant leukocytes have been classified using an ensemble classifier. Only a small subset of the top-ranked features have, however, been used for feature selection due to the use of a statistical feature selection technique. The use of an ensemble classifier also makes the suggested method slower to compute.

Agaian et al. proposed a method in 2014 that robustly segments cells by converting the input RGB image to L*a*b space [27, 33]. They segmented leukocytes using the k-means clustering method. Using Euclidean distance, each pixel in the L*a*b color space had been divided into k clusters. The clusters represent nucleus, background, and other cells. To distinguish between healthy and blast leukocytes, they had also suggested calculating cell energy and Hausdorff dimension (HD). They had investigated shape, color, and texture-based features; however, they have not used a feature selection process to eliminate the extraneous features.

LBP feature had also been used by Singhal et al. in 2014 [34]. To get good classification performance, they examined shape-based features and LBP features. They had not, however, examined other features, like those based on texture and color. Their classification accuracy was not very good either.

Putzu et al. improved the feature-based classification method in 2014 by incorporating shape-based, color-based, and texture-based features [35]. They had also classified healthy and blast leukocytes using an ensemble classifier. To improve classification accuracy, no feature selection mechanism had, however, been used in their work.

Neoh et al. employed the clustering technique in 2015 and used the stimulating discriminant measure (SDM) [36]. Inter and intra cluster variances had been used to evaluate the proposed SDM. In order to cluster the nucleus, cytoplasm, and background region, the SDM had also been utilized in conjunction with GA. A total of 80 features were taken into account in their research; however, no discriminant feature selection methodology was used.

Despite the fact that many researchers have proposed feature-based classification techniques, their classification accuracy is not very high [27–36]. That is due, in part, to the existence of redundant features that have been employed for classification. In order to effectively distinguish between healthy and blast leukocytes, it is essential to choose a relevant group of important features. In this regard, Huang et al. employed PCA to choose 7 discriminant features from a set of 85 features in 2012 [37]. They had classified five different categories of leukocytes using a GA-based k-means clustering strategy. To choose the top-ranked features, Madhloom et al. used a feature ranking method based on Fisher discriminant ratio in 2012 [31]. They only chose 7 features for leukocyte classification out of a total of 30 features. When evaluating the statistical significance of extracted features in 2013, Mohapatra et al. employed a t-test methodology [32]. 32 informative features from a set of 44 features have been selected in their proposed methodology.

Researchers have also suggested evolutionary optimization technique-based feature selection strategies in this context. Osowski et al. suggested a GA-based technique working in conjunction with the Gaussian kernel SVM classifier in 2009 [38]. The features were chosen using GA, and the SVM classifier used the features to identify and classify the cells.

Escalante et al. proposed an ensemble particle swarm model in 2012 for automatic model selection for ALL classification [39]. In their suggested methodology, which really helps in the automatic construction of an ensemble classifier, the search space of candidate classifiers has been investigated. Without human input, their technology may produce an efficient classification model. As a result, the system was more robust to noisy data and yielded more reliable diagnostic predictions.

For the purpose of creating a decision support system for ALL diagnosis, Srisukkham et al. developed a bare-bone PSO-based feature optimization technique in 2017 [40]. The methodology they suggested includes a food-chase and particle-fleeing mechanism. By including an attraction mechanism to food and an avoidance strategy from foes, the search diversity has been improved. The average personal best experience and the global best solution have had an impact on the food-chasing approach. By implementing an enemy avoidance mechanism, the particles were able to migrate away from unproductive search locations.

The gaps that might be found in the reported literature are therefore listed below. It can be seen that by examining several leukocyte properties, the classification performance between healthy cells and blast cells may be further enhanced. Furthermore, it is essential to choose a small number of discriminating features that can distinguish between blast cells and healthy cells with sufficient accuracy. To further demonstrate the effectiveness of the classification capacity, blind testing on the data set is also required.

3.4 Intelligent Computing Approaches for Benchmark Carcinogenic Data Set III

The intelligent diagnosis system also heavily relies on the detection and classification of tumors from tissue level data. Because lung cancer has a high death rate, this thesis focuses on the identification and classification of lung cancer using lung CT images [41, 42]. Pulmonary nodules must be detected since they may be a reliable indication of lung malignancies, making their diagnosis necessary for lung cancer detection [43]. Automated detection of malignant nodules in the lungs may result in the misclassification of benign masses, such as normal anatomical structures, as cancerous. These wrongly classified masses are called false positives (FPs). To avoid this, effective image processing and machine learning techniques must be utilized to ensure accurate identification of malignant lung nodules and to eliminate benign masses. Utilizing effective computational procedures would improve the performance of oncogenic lung nodule detection. Computed tomography (CT) is considered a sensitive imaging modality for accurately detecting pulmonary nodules, whether done manually or computationally [44]. Although CT is an effective imaging modality for detecting pulmonary nodules, the manual process of detection can be tedious and time-consuming for radiologists. Due to the high number of scans required for accurate diagnosis, diagnostic predictions made by radiologists can sometimes be erroneous. Therefore, the development of an automated nodule classification scheme is necessary to assist radiologists with diagnostic interpretation.

Over the past two decades, numerous research studies have focused on the detection and classification of lung nodules. In automated lung nodule detection, most researchers have performed common operations such as lung segmentation, nodule detection, and classification. Various lung segmentation methodologies have been documented in the literature.

In 1998, Kanazawa et al. proposed a lung segmentation technique that uses a simple threshold-based method [45]. Their approach transformed the original image into a binary image using gray level thresholding and used fuzzy clustering to segment blood vessels and nodules within the lung fields. They proposed a gray weighted distance transform technique to extract the region of interest and focused on the shape, gray value, and position of each region to identify it as a blood vessel or nodule. However, this method failed to detect nodules with unclear boundaries and nodules located near blood vessels with diameters less than 4mm.

Armato et al. used a thresholding method to segment the lung structure from the thorax in 1999 [46]. To avoid the loss of juxta-pleural nodules, they proposed a rolling ball algorithm. They also modified the thresholding technique proposed by Kanazawa et al. in [45] by using multiple gray level thresholds for proper identification of lung nodule candidates. Geometric and gray level features were analyzed for each nodule candidate to identify nodule and non-nodule candidates. However, in their study, only 9 features (6 geometric and 3 gray levels) were analyzed for the classification task. More features need to be analyzed to enhance the performance of the classification task.

Hu et al. modified the thresholding technique proposed in [46] in 2001 by introducing an optimal thresholding method that automatically selects a threshold value that reflects the gray level characteristics of lung CT images [47]. They also separated the left and right lungs by identifying the anterior and posterior junctions using a dynamic programming method [47]. Afterward, they smoothed the irregular boundary of the lungs using a morphological operation. However, to segment the lungs, the authors retained only the two largest components of the lung volume that were greater than one percent of the total lung volume.

Li et al. modified the lung segmentation technique in 2003 by introducing a landmark and intensity-based registration algorithm [48]. The proposed algorithm characterizes the mapping between the template image and other image data sets. It also minimizes the inverse consistency error between the template and the target lung volumes. However, the method lacks high segmentation accuracy because finding the proper lung border is challenging.

In 2005, Sluimer et al. proposed an improved registration-based segmentation method to accurately locate the lung border [49]. The method employed an elastic registration scheme with mutual information as a similarity measure. The authors demonstrated that the refinement step led to a significant improvement in segmentation accuracy compared to landmark-based registration techniques. However, the computation time for lung segmentation can be further improved, and the method needs to be validated on a larger set of lung images for better accuracy.

De et al. introduced a method in 2011 for segmenting the lung area using connected component labeling [50]. Their technique involved combining a multipass method with a 3D region growing technique that used a wave-front simulation model. To prevent flooding into the lung parenchyma, the region growing process was interrupted successfully. The starting point for the region growing was automatically chosen by identifying the section of the scan where the lung sections were largest. While their method achieved accurate segmentation, it was unable to segment lungs with various diseases.

Li et al. presented a supervised recursive segmentation scheme in 2015 to reconstruct lungs and vessel trees [51]. Their approach used a supervised semi-3D lung tissue segmentation method to create non-sheltered 3D models of lungs and vessel trees. They proposed a recursive strategy based on geometric active contour to extract lung tissues from CT images. In this strategy, the segmentation of lungs in a given slice was guided by the results of the previous slice, resulting in fast and accurate segmentation of all lung tissues across all CT slices. However, their method had some limitations. First, it was a semi-3D method and required manual intervention. Second, the computational time required was high due to the large data set and manual intervention.

Several methods had been proposed for detecting lung nodules. One such technique was introduced by Lee et al. in 2001, which used a GA-based template matching approach to detect lung nodules within the lung area [52]. In their method, GA was employed to select an efficient template image from multiple reference templates for quick template matching. They also proposed lung wall template matching to detect nodules on the lung wall area using a semi-circular model as a reference pattern. After initial nodule detection using two template matching operations, they evaluated 13 features to eliminate false positive candidates. However, their classification accuracy was not high because their method failed to identify low-contrast nodules, small nodules, and pleural nodules. Additionally, their method produced a high number of false positive candidates.

Ezoe et al. proposed a variable N-quit filter-based method for nodule detection in 2002 [53]. After detection, all the nodule candidates were divided into several clusters using statistical decision theories. The clusters were formed by analyzing nodule sizes and positions in slices, as well as tissue characteristics such as partial volume effect, vertical and non-vertical blood vessels. Within each cluster, nodules were discriminated between normal and abnormal. While their method successfully detected low-contrast nodules, the number of false positive candidates was high.

Li et al. proposed a selective nodule enhancement filter in 2004 to improve nodule detection sensitivity and reduce the number of false positives while suppressing normal anatomic structures such as blood vessels [54]. Their filter was based on eigenvalue analysis of the Hessian matrix and developed selective filters to enhance nodules, vessels, and airways in image slices. The eigenvalues of the Hessian matrix were used to create the filter which was effective in enhancing lung nodules while suppressing other nodule-like anatomical lung structures.

In 2004, Paik et al. proposed a method called surface normal overlap (SNO) for detecting colonic polyps and lung nodules in helical CT images [55]. The SNO score was computed for each voxel within the lung area, based on the number of surface normal passes through it. Nodules and polyps exhibit convex regions on their surfaces, where the inward pointing surface normal vectors tend to intersect. On the other hand, blood vessels also have convex surfaces, but they show a dominant curvature in one direction. As a result, the SNO score for blood vessels is lower compared to nodules and polyps. This SNO score is crucial for distinguishing lung nodules from blood vessels. However, the authors acknowledged a limitation in their work. They did not perform formal optimization for the preprocessing and segmentation tasks, although optimization was conducted for parameters that had minimal impact on the algorithm's performance.

In 2005, Lin et al. introduced a neural network-based fuzzy model for the detection of lung nodules [56]. This model combined the learning and adaptation capabilities of neural networks with a fuzzy inference system to identify pulmonary nodules. The fuzzy inference rules were automatically determined by the proposed method, and the membership function values were refined using the steepest gradient descent algorithm. The neural network-based fuzzy model was effective in classifying nodules from other non-nodule structures. However, the authors focused on a limited set of features for nodule detection, namely the size of the area, mean brightness of the area, and circularity. They did not incorporate surface characteristics and texture features in their methodology, despite their potential usefulness. Additionally, the authors were able to achieve a low computing time for lung segmentation.

In 2011, Pu et al. introduced a shape analysis approach known as the "break-and-repair" strategy for automated segmentation of human lungs [57]. Their method utilized principal curvatures to address problematic regions on the lung surface. They also employed a radial basis function-based implicit surface fitting technique to obtain a comprehensive boundary for the lung surface. However, this method had a limitation in capturing local sharp features of lung regions. Consequently, it may fail to detect pleural nodules located on such sharp regions.

In 2013, Hogeweg et al. presented a methodology for suppressing elongated structures in lung CT images [58]. They utilized a subspace filtering technique to model and reconstruct the elongated structures. A curve aligned with the structure of interest was employed to perform the suppression task. However, the algorithm's effectiveness was found to be reduced when there was a significant variation in the presence of a structure along its centerline. This limitation was particularly observed with the ribs near the chest wall, leading to less effective suppression of elongated structures.

In 2014, Sun et al. proposed a method for segmenting juxta vascular nodules using flow entropy and geodesic distance [59]. Their approach involved evaluating the flow direction feature for each pixel, which was determined based on the orthogonal vector to the gradient vector of all pixels in its neighborhood. The flow direction vector exhibited a low entropy value for pixels within blood vessels compared to those within lung nodules. The authors then employed k-means clustering with geodesic distance to separate the nodules from the blood vessels. However, segmenting large nodules using this method required a significant amount of processing time. Additionally, due to the iterative segmentation and geodesic approach, the overall computing time was lengthy, although it was still within certain limits.

In 2015, Ciompi et al. introduced a descriptor called "bag-of-frequencies" for characterizing pulmonary nodules [60]. This descriptor was employed to encode the information regarding nodule morphology. The intensity profiles of spherical surfaces were sampled to capture the morphology details. Each intensity profile was treated as a periodic signal, and the Fourier transform was used to calculate the Fourier spectrum of each signal. Unsupervised clustering was then performed to label the dictionary of spectra. The resulting descriptor was obtained as a histogram of the labels of the spectra. However, one limitation of this method was the potential increase in the size of the descriptor. To overcome this drawback, the authors suggested utilizing feature selection or feature reduction techniques. These approaches can help mitigate the issue of an excessively large descriptor size.

In 2015, Han et al. utilized a hierarchical vector quantization scheme for precise segmentation of lungs from the chest volume [61]. Additionally, the authors employed rule-based filtering and a feature-based SVM classifier to reduce false positives. While the features used in this work were gathered from previously published studies, the authors acknowledged the potential for further improvement in the nodule classification task by incorporating more sophisticated features, such as texture features.

In 2016, Goncalves et al. proposed a Hessian-based method for segmenting lung nodules [62]. They employed shape index and curvature properties to extract the nodules. The method was extended to a multi-scale approach to accommodate large nodule sizes. Additionally, the authors utilized the central adaptive medialness principle for lung nodule detection [63]. They combined the advantages of both the Hessian-based and central adaptive medialness methods. The central adaptive medialness method enhanced blob-like structures, while the Hessian-based method enhanced small objects and the object borders. This combination resulted in a robust segmentation technique. Although the authors achieved good performance in nodule segmentation, their work did not specifically focus on classifying nodules as malignant or benign.

In 2018, Saien et al. introduced the sparse field level sets and boosting algorithms for lung nodule detection [64]. The sparse field method was used to reconstruct 3D regions of each detected lung nodule. The level set modeling was enhanced by restricting the calculations to a narrow band near the evolving curve, improving computational efficiency. A set of features were extracted for each segmented nodule, and a boosting algorithm was employed to analyze and discriminate nodules from non-nodules based on these features. However, the performance of this method was found to degrade for small-sized nodules, nodules with unusual shapes, and nodules with low contrast.

Additionally, certain feature extraction and classification-based approaches had been applied to the identification of lung nodules. In 2010, Da et al. presented six stages for an automatic lung nodule identification technique, including thorax extraction, lung extraction, lung reconstruction, structures extraction, tubular structures elimination, and false positive reduction [65]. All of the patient's body's artefacts have been thrown away at the thorax extraction step. During the lung extraction stage, the CT image's lung parenchyma was segmented. The lung reconstruction stage has corrected the improper removal of parenchyma related sections. The stage of structure extraction is primarily in charge of choosing the dense structures from within the lung parenchyma. All of the lung's anatomical components other than nodules have been removed during the false positive reduction stage. When the nodule border is not clearly defined, this approach struggles to segment the nodules.

To distinguish between malignant and benign nodules, Chen et al. evaluated morphological criteria like size, margin, contour, and internal characteristics of the nodules in 2012 [66]. The patient's age, gender, and previous history of bloody sputum have also been noted. They have also used ANN and multivariate logistic regression in their work to categorize nodules based on the assessed attributes. Three factors, including calibration, discrimination, and clinical applicability, have been used to compare the effectiveness of ANN and logical regression. The ANN and logical regression, however, have not undergone external validation testing.

In 2013, Sun et al. examined the SVM-based classification for lung cancer detection [67]. To train the classifier and assess its performance, a set of features including morphological traits, patient characteristics, and textural features were used. The performance of the classification process utilizing the Matthews correlation coefficient [68] has also been evaluated using a tenfold cross-validation framework. The usage of the same patient's photos as training samples in order to have a sufficient number of samples for analysis is the study's limitation.

Setio et al. introduced a multi-view convolution network in 2016 in which discriminative features are automatically learnt from the training data [69]. Nodule candidates have been fed into this suggested network. To obtain the classification result in this method, the output of various convolutional networks has been mixed. The pleural nodules and subsolid nodules, however, are not picked up by this approach. So, it is possible to enhance sensitivity performance.

Filho et al. [70] suggested an approach based on shape measuring in 2017. In order to describe spherical things, they created a shape diagram. To evaluate the sphericity of the volume of interest, they have also suggested an entirely new metric. They have also created a cylinder-based method to gain more knowledge about the volume of interest's shape based on exterior geometry. The use of shape metrics is constrained when it comes to analyzing objects with similar shapes but belonging to different classes. As a result, an effective method should be used to solve this issue.

The literature has recently reported on stochastic optimization-based feature selection methodologies for a variety of applications. In order to identify non-technical losses in power distribution systems, Ramos et al. suggested a quick and precise method of feature selection using the HS algorithm in 2011 [71]. They have also suggested an ideal path forest classifier to raise classification accuracy.

Huang et al. applied a self-adaptive HS-based local feature selection technique for music genre categorization in 2014, significantly enhancing the HS-based feature selection task [72]. For handwritten character identification, Stefano et al. suggested a GA-based feature selection strategy this year [73]. To enhance the effectiveness of a face recognition system, Krishna et al. suggested a PSO-based feature selection technique [74]. Inbarani et al. developed a supervised feature selection methodology based on PSO hybridization for the detection of various diseases, such as heart problem, breast cancer, etc. [75].

By combining the idea of rough set rapid reduction with the improved HS algorithm, Inbarani et al. further updated the hybrid feature selection algorithm in 2015 to categorize medical data such as thyroid, lung, and breast cancer [76]. In order to handle the high dimensional feature selection job in large-scale 0–1 knapsack problems, Kong et al. suggested a binary HS technique that year [77]. Additionally, Salam et al. introduced a hybrid feature selection technique for microarray gene expression data in 2016 that makes use of symmetrical uncertainty and the HS idea for gene selection [78].

Although there is still potential for improvement, the previously published lung nodule detection approaches have all attained varying degrees of accuracy. Lung cancer detection therefore places a lot of emphasis on identifying lung nodules and increasing classification accuracy. By examining more features from lung nodules, the performance of lung nodule classification can be enhanced. Additionally, choosing a limited number of significant features is crucial in order to distinguish between healthy and malignant lung nodules and thereby increase classification accuracy.

3.5 Summary

The reported literature in the area of intelligent diagnosis of cancer-causing disorders is summarized in this chapter. Using data at the gene, cell, and tissue levels, it has been discovered that intelligent diagnostic procedures can be created. In order to accurately detect cancerous conditions, this data has been analyzed using the appropriate machine learning algorithms. In addition, for the precise diagnosis of malignancies, classifications of various carcinogenic conditions, healthy and cancerous cells, and malignant and benign tumors have been made in the literature.

The goal of gene level data-based techniques is to identify a small fraction of useful genes that can appropriately categorize various cancer types. This book has concentrated on this problem in gene level data-based analysis. Additionally, some material that has examined various properties from this information has been reported using information-based methodologies at the cell and tissue level. It is a monumental challenge to identify a meaningful subset of informative qualities that would aid in enhancing the effectiveness of intelligent diagnostics. This book has mostly focused on this problem in cell and tissue level-based analysis. These features of intelligent carcinogenic disease identification and classification are covered in more detail in the following chapters.

References

1. Babu M (2004), Computational genomics: Theory and application, chapitre an introduction to microarray data analysis
2. Golub TR, Slonim DK, Tamayo P, Huard C, Gaasenbeek M, Mesirov JP, Coller H, Loh ML, Downing JR, Caligiuri MA et al (1999) Molecular classification of cancer: class discovery and class prediction by gene expression monitoring. Science 286:531
3. Khan J, Wei JS, Ringner M, Saal LH, Ladanyi M, Westermann F, Berthold F, Schwab M, Antonescu CR, Peterson C et al (2001) Classification and diagnostic prediction of cancers using gene expression profiling and artificial neural networks. Nat Med 7:673
4. Hedenfalk I, Duggan D, Chen Y, Radmacher M, Bittner M, Simon R, Meltzer P, Gusterson B, Esteller M, Raffeld M et al (2001) Gene-expression profiles in hereditary breast cance. New England J Med 344:539
5. Tibshirani R, Hastie T, Narasimhan B, Chu G (2002) Diagnosis of multiple cancer types by shrunken centroids of gene expression. Proc Natl Acad Sci 99:6567
6. Guyon I, Weston J, Barnhill S, Vapnik V (2002) Gene selection for cancer classification using support vector machines. Mach Learn 46:389
7. Chu W, Ghahramani Z, Falciani F, Wild DL (2005) Biomarker discovery in microarray gene expression data with Gaussian processes. Bioinformatics 21:3385
8. Fu LM, Fu-Liu CS (2005) Evaluation of gene importance in microarray data based upon probability of selection. BMC Bioinform 6:1
9. Duan K-B, Rajapakse JC, Wang H, Azuaje F (2005) Multiple SVM-RFE for gene selection in cancer classification with expression data. IEEE Trans Nanobiosci 4:228
10. Yang K, Cai Z, Li J, Lin G (2006) A stable gene selection in microarray data analysis. BMC Bioinform 7:1
11. Kestler HA, Müssel C 2006 An empirical comparison of feature reduction methods in the context of microarray data classification. In: Artificial Neural Networks in Pattern Recognition: Second IAPR Workshop, ANNPR 2006, Ulm, Germany, August 31–September 2, 2006. Proceedings 2. Springer, Berlin, pp. 260–273
12. Tang Y, Zhang Y-Q, Huang Z (2007) Development of two-stage SVM-RFE gene selection strategy for microarray expression data analysis. IEEE/ACM Trans Comput Biol Bioinform 4:365
13. Niijima S, Kuhara S (2006) Recursive gene selection based on maximum margin criterion: a comparison with SVM-RFE. BMC Bioinform 7:1
14. Kittler J et al (1978) Pattern recognition and signal processing. In: Chapter Feature Set Search Algorithms Sijthoff and Noordhoff, Alphen aan den Rijn, Netherlands, pp 41–60
15. Pudil P, Novovičová J, Kittler J (1994) Floating search methods in feature selection. Pattern Recogn Lett 15:1119
16. Kohavi R, John GH (1997) Wrappers for feature subset selection. Artif Intell 97:273
17. Inza I, Sierra B, Blanco R, Larrañaga P (2002) Gene selection by sequential search wrapper approaches in microarray cancer class prediction. J Intell Fuzzy Syst 12:25
18. Zhu Z, Ong Y-S, Dash M (2007) Markov blanket-embedded genetic algorithm for gene selection. Pattern Recogn 40:3236
19. Koller D, Sahami M et al (1996) Toward optimal feature selection. In: ICML, vol 96, p 292
20. Li J, Su H, Chen H, Futscher BW (2007) Optimal search-based gene subset selection for gene array cancer classification. IEEE Trans Inf Technol Biomed 11:398
21. Shen Q, Shi W-M, Kong W (2008) Hybrid particle swarm optimization and tabu search approach for selecting genes for tumor classification using gene expression data. Comput Biol Chem 32:53
22. Chu F, Jin G, Wang L (2005) Cancer diagnosis and protein secondary structure prediction using support vector machines. In: Support vector machines: Theory and applications. Springer, pp 343–363

23. Li S, Wu X, Tan M (2008) Gene selection using hybrid particle swarm optimization and genetic algorithm. Soft Comput 12:1039
24. Mohamad MS, Omatu S, Deris S, Yoshioka M (2011) A modified binary particle swarm optimization for selecting the small subset of informative genes from gene expression data. IEEE Trans Inf Technol Biomed 15:813
25. Li H, Nantasenamat C, Monnor T, Isarankura-Na-Ayudhya C, Prachayasittikul V (2013) Genetic algorithm search space splicing particle swarm optimization as general-purpose optimizer. Chemom Intell Laborat Syst 128:153
26. Chen K-H, Wang K-J, Tsai M-L, Wang K-M, Adrian AM, Cheng W-C, Yang T-S, Teng N-C, Tan K-P, Chang K-S (2014) Gene selection for cancer identification: a decision tree model empowered by particle swarm optimization algorithm. BMC Bioinf 15:1
27. Sinha N, Ramakrishnan A (2003) Automation of differential blood count. In: TENCON 2003. Conference on Convergent Technologies for Asia-Pacific Region, vol 2. IEEE, Piscataway, pp 547–551
28. Piuri V, Scotti F (2004) Morphological classification of blood leucocytes by microscope images. In: 2004 IEEE International Conference onComputational Intelligence for Measurement Systems and Applications, 2004. CIMSA. IEEE, Piscataway, pp 103–108
29. Scotti F (2006) Robust segmentation and measurements techniques of white cells in blood microscope images. In: 2006 IEEE instrumentation and Measurement Technology Conference Proceedings. IEEE, Piscataway, pp 43–48
30. Rezatofighi SH, Soltanian-Zadeh H (2011) Automatic recognition of five types of white blood cells in peripheral blood. Comput Med Imag Graph 35:333
31. Madhloom HT, Kareem SA, Ariffin H (2012) A robust feature extraction and selection method for the recognition of lymphocytes versus acute lymphoblastic leukemia. In: 2012 International Conference on Advanced Computer Science Applications and Technologies (ACSAT). IEEE, Piscataway, pp 330–335
32. Mohapatra S, Patra D, Satpathy S (2014) An ensemble classifier system for early diagnosis of acute lymphoblastic leukemia in blood microscopic images. Neural Comput Appl 24:1887
33. Agaian S, Madhukar M, Chronopoulos AT (2014) Automated screening system for acute myelogenous leukemia detection in blood microscopic images. IEEE Syst J 8:995
34. Singhal V, Singh P (2014) Local binary pattern for automatic detection of acute lymphoblastic leukemia. In: 2014 Twentieth National Conference on Communications (NCC). IEEE, Piscataway, pp 1–5
35. Putzu L, Caocci G, Di Ruberto C (2014) Leucocyte classification for leukaemia detection using image processing techniques. Artif Intell Med 62:179
36. Chin Neoh S, Srisukkham W, Zhang L, Todryk S, GreystokeB, Peng Lim C, Alamgir Hossain M, Aslam N (2015) An intelligent decision support system for leukaemia diagnosis using microscopic blood images. Sci Rep 5:1
37. Huang D-C, Hung K-D (2012) Leukocyte nucleus segmentation and recognition in color blood-smear images. In: 2012 IEEE International Instrumentation and Measurement Technology Conference Proceedings. IEEE, Piscataway, pp 171–176
38. Osowski S, Siroic R, Markiewicz T, Siwek K (2008) Application of support vector machine and genetic algorithm for improved blood cell recognition. IEEE Trans Instrument Measure 58:2159
39. Escalante HJ, Montes-y Gómez M, González JA, Gómez-Gil P, Altamirano L, Reyes CA, Reta C, Rosales A (2012) Acute leukemia classification by ensemble particle swarm model selection. Artif Intell Med 55:163
40. Srisukkham W, Zhang L, Neoh SC, Todryk S, Lim CP (2017) Intelligent leukaemia diagnosis with bare-bones PSO based feature optimization. Appl Soft Comput 56:405
41. De Valk J, Eijkman E (1984) Analysis of eye fixations during the diagnostic interpretation of chest radiographs. Med Biol Eng Comput 22:353
42. Greenlee RT, Murray T, Bolden S, Wingo PA (2000) Cancer statistics, 2000. CA: A Cancer J Clinic 50:7

43. Chen S, Suzuki K (2012) Computerized detection of lung nodules by means of "virtual dual-energy" radiography. IEEE Trans Biomed Eng 60:369
44. Henschke CI, McCauley DI, Yankelevitz DF, Naidich DP, McGuinness G, Miettinen OS, Libby DM, Pasmantier MW, Koizumi J, Altorki NK et al (1999) Early Lung Cancer Action Project: overall design and findings from baseline screening. Lancet 354:99
45. Kanazawa K, Kawata Y, Niki N, Satoh H, Ohmatsu H, Kakinuma R, Kaneko M, Moriyama N, Eguchi K (1998) Computer-aided diagnosis for pulmonary nodules based on helical CT images. Comput Med Imag Graph 22:157
46. Armato SG, Giger ML, Moran CJ, Blackburn JT, Doi K, MacMahon H (1999) Computerized detection of pulmonary nodules on CT scans. Radiographics 19:1303
47. Hu S, Hoffman EA, Reinhardt JM (2001) Automatic lung segmentation for accurate quantitation of volumetric X-ray CT images. IEEE Trans Med Imag 20:490
48. Li B, Christensen GE, Hoffman EA, McLennan G, Reinhardt JM (2003) Establishing a normative atlas of the human lung: intersubject warping and registration of volumetric CT images. Acad Radiol 10:255
49. Sluimer I, Prokop M, Van Ginneken B (2005) Toward automated segmentation of the pathological lung in CT. IEEE Trans Med Imag 24:1025
50. De Nunzio G, Tommasi E, Agrusti A, Cataldo R, De Mitri I, Favetta M, Maglio S, Massafra A, Quarta M, Torsello M et al (2011) Automatic lung segmentation in CT images with accurate handling of the hilar region. J Digital Imag 24:11
51. Li X, Wang X, Dai Y, Zhang P (2015) Supervised recursive segmentation of volumetric CT images for 3D reconstruction of lung and vessel tree. Comput Methods Programs Biomed 122:316
52. Lee Y, Hara T, Fujita H, Itoh S, Ishigaki T (2001) Automated detection of pulmonary nodules in helical CT images based on an improved template-matching technique. IEEE Trans Med Imag 20:595
53. Ezoe T, Takizawa H, Yamamoto S, Shimizu A, Matsumoto T, Tateno Y, Iimura T, Matsumoto M (2002) Automatic detection method of lung cancers including ground-glass opacities from chest X-ray CT images: In: Medical Imaging 2002: Image Processing, vol 4684 . SPIE, Bellingham, pp 1672–1680
54. Li Q, Doi K (2004) Medical Imaging 2004: Image Processing. In: Medical Imaging 2004: Image Processing, vol 5370. SPIE, Bellingham, pp 1–9
55. Paik DS, Beaulieu CF, Rubin GD, Acar B, Jeffrey RB, Yee J, Dey J, Napel S (2004) Surface normal overlap: a computer-aided detection algorithm with application to colonic polyps and lung nodules in helical CT. IEEE Trans Med Imag 23:661
56. Lin D-T, Yan C-R, Chen W-T (2005) Autonomous detection of pulmonary nodules on CT images with a neural network-based fuzzy system. Comput Med Imag Graph 29:447
57. Pu J, Paik DS, Meng X, Roos J, Rubin GD (2010) Shape "break-and-repair" strategy and its application to automated medical image segmentation. IEEE Trans Visual Comp Graph 17:115
58. Hogeweg L, Sánchez CI, van Ginneken B (2013) Suppression of translucent elongated structures: applications in chest radiography. IEEE Trans Med Imag 32:2099
59. Sun S, Guo Y, Guan Y, Ren H, Fan L, Kang Y (2014) Juxta-vascular nodule segmentation based on flow entropy and geodesic distance. IEEE J Biomed Health Inf 18:1355
60. Ciompi F, Jacobs C, Scholten ET, Wille MM, De Jong PA, Prokop M, van Ginneken B (2014) Bag-of-frequencies: a descriptor of pulmonary nodules in computed tomography images. IEEE Trans Med Imag 34:962
61. Han H, Li L, Han F, Song B, Moore W, Liang Z (2014) Fast and adaptive detection of pulmonary nodules in thoracic CT images using a hierarchical vector quantization scheme. IEEE J Biomed Health Inf 19:648
62. Gonçalves L, Novo J, Campilho A (2016) Hessian based approaches for 3D lung nodule segmentation. Expert Syst Appl 61:1
63. Novo J, Gonçalves L, Mendonça AM, Campilho A (2015) 3D lung nodule candidate detection in multiple scales. In: 2015 14th IAPR International Conference on Machine Vision Applications (MVA). IEEE, Piscataway, pp 61–64

64. Saien S, Moghaddam HA, Fathian M (2018) A unified methodology based on sparse field level sets and boosting algorithms for false positives reduction in lung nodules detection. Int J Comput Assist Radiol Surgery 13:397
65. da Silva Sousa JRF, Silva AC, de Paiva AC, Nunes RA (2010) Methodology for automatic detection of lung nodules in computerized tomography images. Comput Methods Program Biomed 98:1
66. Chen H, Zhang J, Xu Y, Chen B, Zhang K (2012) Performance comparison of artificial neural network and logistic regression model for differentiating lung nodules on CT scans. Expert Syst Appl 39:11503
67. Sun T, Wang J, Li X, Lv P, Liu F, Luo Y, Gao Q, Zhu H, Guo X (2013) Comparative evaluation of support vector machines for computer aided diagnosis of lung cancer in CT based on a multi-dimensional data set. Comput Methods Progr Biomed 111:519
68. Matthews BW (1975) Comparison of the predicted and observed secondary structure of T4 phage lysozyme. Biochimica et Biophysica Acta (BBA)-Protein Structure 405:442
69. Setio AAA, Ciompi F, Litjens G, Gerke P, Jacobs C, Van Riel SJ, Wille MMW, Naqibullah M, Sánchez CI, Van Ginneken B (2016) Pulmonary nodule detection in CT images: false positive reduction using multi-view convolutional networks. IEEE Trans Med Imag 35:1160
70. Filho AOdC, Silva AC, de Paiva AC, Nunes RA, Gattass M (2017) 3D shape analysis to reduce false positives for lung nodule detection systems. Med Biol Eng Comput 55:1199
71. Ramos CC, Souza AN, Chiachia G, Falcão AX, Papa JP (2011) A novel algorithm for feature selection using harmony search and its application for non-technical losses detection. Comput Electri Eng 37:886
72. Huang Y-F, Lin S-M, Wu H-Y, Li Y-S (2014) Music genre classification based on local feature selection using a self-adaptive harmony search algorithm. Data Knowl Eng 92:60
73. De Stefano C, Fontanella F, Marrocco C, Di Freca AS (2014) A GA-based feature selection approach with an application to handwritten character recognition. Pattern Recogn Lett 35:130
74. Krisshna NA, Deepak VK, Manikantan K, Ramachandran S (2014) Face recognition using transform domain feature extraction and PSO-based feature selection. Appl Soft Comput 22:141
75. Inbarani HH, Azar AT, Jothi G (2014) Supervised hybrid feature selection based on PSO and rough sets for medical diagnosis. Comput Methods Progr Biomed 113:175
76. Inbarani HH, Bagyamathi M, Azar AT (2015) A novel hybrid feature selection method based on rough set and improved harmony search. Neural Comput Appl 26:1859
77. Kong X, Gao L, Ouyang H, Li S (2015) A simplified binary harmony search algorithm for large scale 0–1 knapsack problems. Expert Syst Appl 42:5337
78. Shreem SS, Abdullah S, Nazri MZA (2016) Hybrid feature selection algorithm using symmetrical uncertainty and a harmony search algorithm. Int J Syst Sci 47:1312

Chapter 4
Classical Approaches in Gene Evaluation for Carcinogenic Disease Detection

Abstract This chapter explores the critical role that microarray technology plays in computational biology, providing a powerful tool for detecting anomalies in the human body through gene expression information. Microarray data sets are critical to the automated detection of carcinogenic diseases, as they facilitate manual pathological diagnosis methods. The challenges arise from the vast number of genes within limited accessible samples, making the analysis of gene expression levels intricate. Two classical gene selection strategies, namely filter and wrapper methodologies, are explored in this chapter. In filter techniques, genes are ranked statistically, while in wrapper approaches, genes are repeatedly chosen depending on classification outcomes. Although they are predicted to perform better than filter methods, wrapper approaches have a higher computational cost. With a focus on SRBCT, ALL_AML, and MLL subclasses of cancer, the chapter seeks to distinguish cancer types based on patterns of gene expression. For a diagnosis to be both accurate and efficient, filter- and wrapper-based gene selection techniques have been examined. The chapter delves more into gene ranking techniques based on filter methods with binary and multi-class data sets. Additionally, it elaborates on the utility of Particle Swarm Optimization (PSO) methodology in wrapper-based gene selection from microarray gene expression data, contributing to the understanding of effective diagnostic methodologies.

4.1 Introduction

One of the key developments in the area of experimental biology is microarray technology. Gene microarray information has provided the scientific community a tool that can be utilized to find any potential anomalies in the human body. The community working on automated carcinogenic disease identification to support manual pathological diagnostic procedure heavily relies on pertinent data based on the microarray data set.

Due to the existence of a large number of genes in a limited number of accessible samples, it is challenging to analyze these gene expression levels. Therefore, the

K. Das Sharma et al., *Intelligent Computing in Carcinogenic Disease Detection*, Computational Intelligence Methods and Applications,
https://doi.org/10.1007/978-981-97-2424-6_4

data mining community has considerable difficulties in identifying important genes from microarray gene expression data. As previously mentioned, there are two main strategies for gene selection: the filter method-based strategy and the wrapper method-based strategy. In filter-based method, statistical ranking tests are performed on the data set [1–8]. Genes are organized according to their importance using such gene ranking approaches [1]. Following that, a suitable subset of genes is chosen depending on the ranking. In contrast, wrapper-based approaches undertake gene selection iteratively with the aid of classification results, which necessitates the selection of a group of important genes [2–4, 8]. On the basis of several learning techniques, the classifier selects genes. Typically, wrapper-based methods are anticipated to outperform filter-based methods, although at the expense of increased computational overhead [8–10].

In this chapter, we have made an effort to identify three subclasses of cancer: SRBCT, ALL, and MLL. We tried to identify the specific type of cancer using gene expression patterns in order to accomplish our goal. In this context, examining filter-based and wrapper-based gene selection strategies is crucial because they both play specific roles in an efficient and accurate diagnosis.

This chapter explores various filter method-based gene ranking approaches using binary and multi-class data sets [11]. In addition, the utility of the PSO methodology in the wrapper-based method for the selection of genes from microarray gene expression data is further elaborated in this chapter.

4.2 Filter Approaches in Gene Evaluation

Filter-based approaches require a number of stages for choosing significant genes [1]. Each of the genes is given a rank in the first stage. All of the genes are rearranged in the second phase in accordance with their computed scores. The gene ranking list starts with the gene with the highest score, then moves on to the gene with the second-highest score, and so forth. Next, to determine classification accuracy, the ordered genes with higher scores are used successively. The block diagram of the filter method-based gene selection technique is shown in Fig. 4.1.

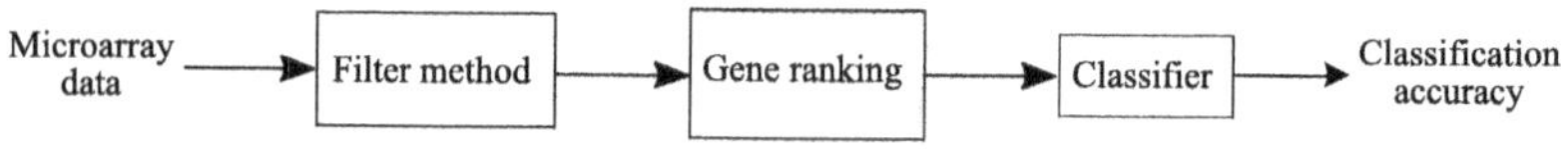

Fig. 4.1 Filter method-based gene selection technique

4.2.1 T-Test Technique

The t-distribution is a probability distribution that is typically employed to estimate the population parameters when the sample size is small. However, statisticians rely on the distribution of t-statistic when sample sizes are small [12].

A statistical hypothesis often describes certain population distribution parameters. The hypothesis is accepted if a sample from the population distribution supports it; otherwise, the hypothesis is rejected [12]. Typically, the null hypothesis H_0 is used to denote the hypothesis. For example, if a population have normal distribution with mean μ, then the hypothesis about μ is $H_0 : \mu = 1$. This particular hypothesis suggests that the population is normal, with a mean of 1. Similarly, all samples from a population with n samples should be examined before deciding whether or not to accept the null hypothesis H_0. A test statistic is a random variable used in a hypothesis test to decide whether to accept or reject the null hypothesis.

Case-I Assume a population of n number of samples, $x_1, x_2, \ldots, x_n$, have unknown mean μ and known variance σ^2. To test the null hypothesis $H_0 : \mu = \mu_0$, where μ_0 is some user specified constant, the mean of all the samples is calculated. The mean value is calculated by $\overline{x} = \sum_{i=1}^{n} \frac{x_i}{n}$, where x_i is the ith number of sample and $\overline{x}$ is the mean value. The hypothesis H_0 is accepted if $\overline{x}$ is close to μ_0. However, when $\mu = \mu_0$, $\overline{x}$ is distributed normally with mean μ_0 and variance $\frac{\sigma^2}{n}$ and a standard normal random variable z is defined by $z = \frac{\overline{x}-\mu_0}{\sigma/\sqrt{n}}$.

Case-II For the same population of case-I having unknown mean μ and unknown variance σ^2, the same hypothesis is tested. The hypothesis H_0 is rejected if the sample mean $\overline{x}$ is far from μ_0. However, how distant it should be to confirm rejection is dependent on σ^2. In this case, σ^2 is estimated by $s^2 = \frac{\sum_{i=1}^{n}(x_i-\overline{x})^2}{n-1}$, and the hypothesis is rejected when $\left|\frac{\overline{x}-\mu}{s/\sqrt{n}}\right|$ is large. Therefore, the statistics T, expressed by $T = \frac{\sqrt{n}(\overline{x}-\mu_0)}{s}$, has a t-distribution when $\mu = \mu_0$.

Case-III Let us consider two populations, $x_1, x_2, \ldots, x_n$ and $y_1, y_2, \ldots, y_m$ having unknown parameters (μ_x, σ_x^2) and $\left(\mu_y, \sigma_y^2\right)$, and the hypothesis $H_0 : \mu_x = \mu_y$ is to be tested. The hypothesis H_0 is rejected if $\overline{x} - \overline{y}$ is far from zero. However, how distant it should be to confirm rejection σ_x^2 and σ_x^2 is estimated by $s_x^2 = \frac{\sum_{i=1}^{n}(x_i-\overline{x})^2}{n-1}$ and $s_y^2 = \frac{\sum_{i=1}^{m}(y_i-\overline{y})^2}{m-1}$, respectively (assuming $\sigma_x^2 = \sigma_y^2 = \sigma^2$). Therefore, if the hypothesis H_0 is true and $\mu_x - \mu_y = 0$, the statistics $T = \frac{\overline{x}-\overline{y}}{\sqrt{s_p^2\left(\frac{1}{n}+\frac{1}{m}\right)}}$ is a t-distribution, where s_p^2 is the pooled estimator of the common variance σ^2 and is expressed by $s_p^2 = \frac{(n-1)s_x^2+(m-1)s_y^2}{n+m-2}$. This t-statistics is the basic statistical technique for determining the relationship between two differentially expressed groups of data when they have equal sample variances [13].

Microarray gene expression samples, however, do not all have the same variances. As a result, Welch suggested a modification to the common t-statistic for samples with different variances [13]. Welch t-statistics is given by $T = \frac{\overline{x}-\overline{y}}{\sqrt{\frac{s_x^2}{n}+\frac{s_y^2}{m}}}$, where s_x and s_y are the variances of the samples, respectively.

To order the genes in microarray gene expression profiles according to their significance, the modified Welch t-statistics [14] is most frequently utilized. The value of T measured at the n expression levels of the gene is defined as the discriminating power of a gene for a test statistic T. This definition is based on the fact that with high value of T, the null hypothesis $H_0 : \mu_1 = \mu_2 = \ldots = \mu_n$ is likely to be rejected. Therefore, the greater the gene's ability to distinguish between various cancer kinds, the higher its discriminating power is. Finally, a group of informative genes that are highly effective at differentiating between various cancer subtypes must be chosen.

The T-score (TS) for every gene on the microarray is first determined. According to their TS value, all the genes are then rearranged in descending order. The definition of TS for the i_{th} gene can be expressed mathematically as

$$TS_i = max\left\{\left|\frac{\overline{x}_{ik}-\overline{x}_i}{\sqrt{s_i^2\,(1/n_k+1/n)}}\right|, k = 1, 2, \ldots, K\right\} \tag{4.1}$$

where

$$\overline{x}_{ik} = \sum_{j\in C_k} \overline{x}_{ij}/n_k \tag{4.2}$$

$$\overline{x}_i = \sum_{j=1}^{n} x_{ij}/n \tag{4.3}$$

$$s_i^2 = \frac{1}{n-K}\sum_{k}\sum_{j\in C_k}\left(x_{ij}-\overline{x}_{ik}\right)^2 \tag{4.4}$$

K denotes the total number of classes. C_k refers to class k that includes n_k samples; x_{ij} is the expression value of gene i in sample j; $\overline{x}_{ik}$ denotes the mean expression value in k_{th} class for gene i; n is the total number of samples; $\overline{x}_i$ denotes the overall mean expression value for gene i; s_i denotes the within-class standard deviation for gene i.

4.2.2 Chi-Square Technique

Pearson first introduced the chi-squared (χ^2) statistics [15]. The entire data set is not expressed as a single quantity like mean or standard deviation in the traditional chi-squared method. Instead, the value of χ^2 statistics relative to the class is used to calculate each gene's relevance. χ^2 presents the significant difference between two genes, first by determining the distribution of observations (frequencies) assuming no difference exists and second by measuring the extent to which the observed distribution deviates from that determined in the first step. The frequencies of the sample distribution are known as the observed frequencies, while the frequencies assuming no difference occurs are known as expected frequencies. The hypothesis H_0 is the assumption that no difference exists between two genes, and it is tested by the χ^2 statistics. The foundation of this statistics is the difference between the observed and expected frequencies. For each sample of each class, this variation has been squared and then divided by the expected value. The results of adding the numbers for each class are then known as the chi-squared value. The chi-squared statistics has been defined as

$$\chi^2 = \sum_i \frac{(O_i - E_i)^2}{E_i} \tag{4.5}$$

where O_i is the observed frequency of i_{th}class and E_i is the expected frequency of i_{th} class. When there is no difference between two genes, as suggested by the data distribution, the values of observed and expected frequencies are equal. On the other hand, the observed and expected frequencies diverge more as the difference increases. As a result, the gene is more effective in differentiating between two classes the higher the χ^2 value.

This method has been used in the diagnostic prediction of cancers by Liu et al. [16]. By assessing each gene's chi-squared statistics in relation to the classes, the method assesses each gene separately. The gene expression values for each gene are separated into a number of intervals using this procedure. The entire sample size for each gene has been divided into two intervals in this book using the mean expression value of the samples for each gene. Thereafter, the χ^2 value for each gene has been calculated as

$$\chi^2 = \sum_{i=1}^{m} \sum_{j=1}^{k} \frac{\left(A_{ij} - E_{ij}\right)^2}{E_{ij}} \tag{4.6}$$

where

$$E_{ij} = \frac{R_i \times C_j}{N} \tag{4.7}$$

$$R_i = \sum_{j=1}^{k} A_{ij} \tag{4.8}$$

$$C_j = \sum_{i=1}^{m} A_{ij} \tag{4.9}$$

$$N = \sum_{i=1}^{m} R_i = \sum_{i=1}^{m} \sum_{j=1}^{k} A_{ij} \tag{4.10}$$

Here, m denotes the number of intervals. A_{ij} refers to the number of samples in the i_{th} interval, j_{th} class. k is the number of classes. E_{ij} is the expected frequency of A_{ij}. R_i is the number of samples in the i_{th} interval. C_j is the number of samples in the j_{th} class. N denotes the total number of samples.

Similar to the t-test procedure, a rank list array has been created once all the gene's χ^2 values have been determined, with the gene having the highest χ^2 value at the top.

4.2.3 Signal-to-Noise Ratio Technique

The signal-to-noise ratio (SNR) technique [17] finds the expression patterns with a minimum fluctuation of expression within each class and a maximum difference in mean expression between two classes. In the SNR technique, the mean values of the samples from the two classes were first subtracted, and then their respective standard deviation differences were divided. Therefore, a high SNR value denotes a gene's ability to distinguish between two groups.

The difference between the class means, which determines the noise in the microarray gene expression profile, is used to judge the strength of the spot signal. Because it can distinguish between different cancer types, this method has the efficacy of a gene. The mean and standard deviation values of the samples in one class as well as the mean and standard values of the samples in other classes are taken into account when calculating the SNR values for the multi-class problem. This method also rates all the genes in the microarray domain in order to determine how well a gene can distinguish between various classes [17]. The signal-to-noise score for each gene can be calculated as

$$S(g) = \sum_{c}^{i=1} \frac{|m_i - m_i'|}{s_i + s_i'} \tag{4.11}$$

where c is the number of classes, m_i is the mean value of the samples for i_{th} category, m_i' is the mean value of the samples in the other classes, s_i is the standard deviation of the samples in the i_{th} category, and s_i' is the standard deviation of the samples in the other classes. As the more informative gene has a higher SNR value, all the signal-to-noise scores are then arranged in descending order.

According to a study on the filter method-based gene selection strategy, the ranking of useful cancer-specific genes is crucial for the accurate subclassification of cancers. Some of the top rank genes, nevertheless, might be redundant. No extra information for subgroup categorization is carried by the redundant genes. So, in this chapter, the limited subset of meaningful genes from the microarray gene expression profile have been identified using the wrapper method-based gene selection strategy.

4.3 Wrapper Approaches in Gene Evaluation

The performance of the classifier that gives us a collection of distinct classes of genes with distinctly different properties is used in the wrapper-based method to evaluate the performance of the important subset of genes [18, 19]. The wrapper technique starts a search for a useful subset of genes using an effective search tool that can operate over the entire collection of genes. The performance of the chosen subset is then assessed using an appropriate classifier. Depending on the effectiveness of the classification, the evaluation is repeated for each subset. It is needless to say that the search strategy is entirely responsible for the subset generation in the wrapper function.

Because of the aforementioned factors, it is obvious that wrapper approaches would be slower than filter methods in identifying a suitable subset of genes. However, compared to filter approaches, wrapper methods yield subsets with greater classification performance [20].

Therefore, wrapper methods can largely benefit from effective global search approaches. Evolutionary computing has received a lot of interest recently as a global search method for resolving the gene selection issue [21]. GA has recently been applied to microarray gene expression data to pick genes [22–27]. The GA-based techniques, however, involve numerous evolutions. Again, the wrapper technique requires a lot of work for each such evolution. Consequently, the GA-based techniques require a lot of computing time. PSO, on the other hand, has an organized neighborhood that makes the search process easier. PSO also includes a velocity term that permits quick convergence to a solution. Due to its straightforward mathematical description, PSO so clearly outperforms GA. Additionally, it is computationally more effective in terms of memory usage and calculation time [28].

PSO-based wrapper techniques for gene selection have grown in acceptance in recent years [2, 4, 29–32]. The block diagram of wrapper method-based gene selection is shown in Fig. 4.2.

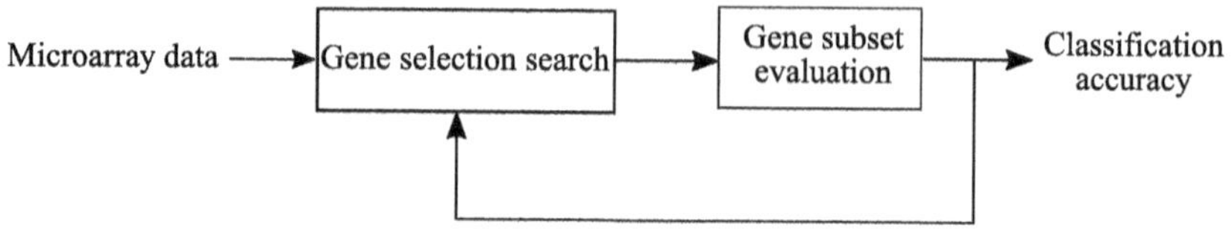

Fig. 4.2 Wrapper method-based gene selection technique

4.3.1 Particle Swarm Optimization (PSO)

PSO is a population-based evolutionary optimization technique developed by Kennedy and Eberhart in 1995 [28]. It is a stochastic optimization process that rests on swarm, where each element of the swarm is known as particle. PSO actually simulates the dynamic behavior of birds in a bird flock or fishes in a school of fish that involves sharing of information and grant particles to take benefit from explorations and previous involvements in the hunt for food. In PSO, each particle represents a candidate solution of the optimization problem. The particles in PSO retain their own optimal position. The particles also retain the optimal position of the swarm. In each iteration, the information of each particle (i.e., own optimal position and swarm's optimal position) is utilized to tune the velocity of each particle. The velocity information is further used to evaluate the new position of the particles. The particles continuously update their information (i.e., position and velocity) in each iteration in the multidimensional search space until they reach in an optimal solution.

Assume there are N number of particles in a swarm. The position and velocity of each particle are d dimensional and represented as $p_i = (p_{i1}, p_{i2}, p_{i3}, \ldots, p_{id})$ and $v_i = (v_{i1}, v_{i2}, v_{i3}, \ldots, v_{id})$. The initial position and velocity of each particle are calculated as

$$p_i^0 = s_{min} + r_1 (s_{max} - s_{min}) \tag{4.12}$$

$$v_i^0 = v_{min} + r_2 (v_{max} - v_{min}) \tag{4.13}$$

where r_1,r_2 are random numbers between 0 and 1 and s_{min}, v_{min}, s_{max}, and v_{max} are the vectors of lower and upper bounds of the particle's position and velocity, respectively.

Each particle keeps track of the best position visited by itself using its own knowledge, and the position is known as personal best (p_{best}) position. Each particle also keeps tracks of the best position obtained so far by any particle. This overall best position is known as the global best (g_{best}) position of the particle. How close a particle is to the g_{best} position is measured by a fitness function, which varies depending upon the optimization problem. The personal best (p_{best}) and global best

(g_{best}) solutions for each particle are calculated as

$$p_{best}(\Delta t+1)=\begin{cases}p_i(\Delta t), & \text{if } FF(p_i(\Delta t+1)) \geq FF(p_i(\Delta t))\\ p_i(\Delta t+1), & \text{if } FF(p_i(\Delta t+1)) < FF(p_i(\Delta t)).\end{cases} \tag{4.14}$$

$$g_{best}(\Delta t)=min\{FF(p_1(\Delta t)), FF(p_2(\Delta t)), \ldots, FF(p_N(\Delta t))\} \tag{4.15}$$

where FF is the fitness function and Δt is the time step. At each time step Δt, the new velocity of each particle is calculated using the information of p_{best} and g_{best} positions.

$$v_i(\Delta t+1)=\omega_t \times v_i(\Delta t)+c_1r_1(p_{best}(\Delta t)-p_i(\Delta t))+c_2r_2(g_{best}(\Delta t) - p_i(\Delta t)) \tag{4.16}$$

The velocity update equation represents that the new velocity, $v_i(\Delta t+1)$, depends upon the previous velocity of the particle, $v_i(\Delta t)$, and executes inertial movement due to influence of the inertia weight, ω_t. The inertia weight in each iteration is calculated as follows:

$$\omega_t=\frac{(iter_{max}-\Delta t)\times(\omega_{start}-\omega_{end})}{iter_{max}}+\omega_{end} \tag{4.17}$$

where $iter_{max}$ is the maximum number of iterations of PSO. The value of inertia weight is decreased from ω_{start} to ω_{end} throughout the progress of PSO iterations [33, 34]. The new velocity of the particle also depends on the distance between personal best position (p_{best}) and current position of the particle. This is known as cognitive part of the velocity update equation (the second term in Eq. 4.16). It demonstrates the particle's movement resulting from its own experience. Hence, the constant term c_1 associated with it is termed as cognitive learning constant or cognitive acceleration constant. The new velocity of the particle also depends on the difference between global best position (g_{best}) and the current position of the particle. This is known as social part of the velocity update equation (the third term in Eq. 4.16). It demonstrates the co-operating and information sharing nature of the particle in the population. Hence, the constant term c_2 associated with it is termed as social learning constant or social acceleration constant.

The new position of each particle is calculated using the information of new velocity as

$$p_i(\Delta t+1)=p_i(\Delta t)+v_i(\Delta t+1) \tag{4.18}$$

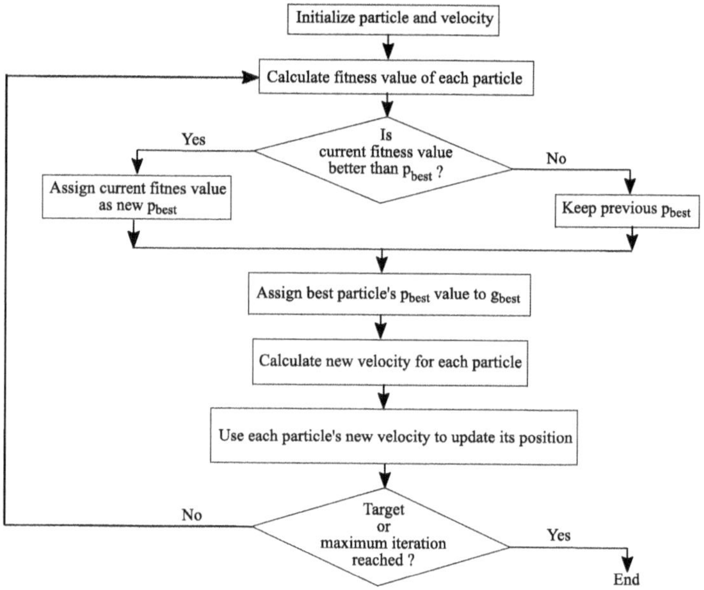

Fig. 4.3 Flowchart of PSO technique

The velocity of each particle, in each dimension, is clamped to a maximum velocity v_{max}, as

$$\text{if} \quad v_i\,(\Delta t + 1) > v_{max}, \text{ then} \quad v_i\,(\Delta t + 1) = v_{max} \tag{4.19}$$

$$\text{if} \quad v_i\,(\Delta t + 1) < -v_{max}, \text{ then} \quad v_i\,(\Delta t + 1) = -v_{max} \tag{4.20}$$

The algorithm is presented in Algorithm 4.1, and the flowchart as depicted in Fig. 4.3 throw lights on PSO technique.

4.3.2 PSO-Based Wrapper Method for Gene Selection

Each PSO particle in this approach stands for a candidate gene subset. The fitness function is then assessed using each candidate gene subset. Following then, each particle has its position and velocity updated based on its p_{best} and g_{best} values. And last, a classification algorithm employs the chosen collection of informative genes to deliver training and testing accuracy [35]. The block diagram of the PSO-based wrapper method for the selection of important gene subset is shown in Fig. 4.4.

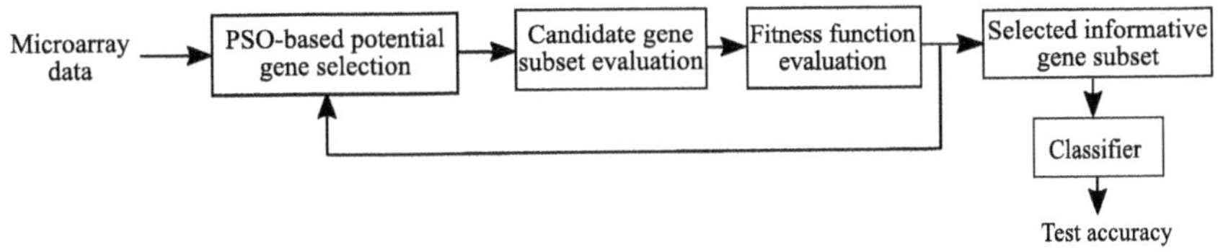

Fig. 4.4 PSO-based wrapper method

Algorithm 4.1 PSO algorithm

for each particle **do**
  initialize position
  initialize velocity
end for
Do until maximum iteration or minimum error criteria reached
for each particle **do**
  calculate fitness value
  if the fitness value is better than p_{best} **then**
    set p_{best}=current fitness value
  end if
  if p_{best} is better than g_{best} **then**
    $g_{best} = p_{best}$
  end if
end for
for each particle **do**
  calculate the new velocity of the particle
  calculate the new position of the particle
end for

The position of each particle is first initialized by generating and associating random numbers between 0 and 1. Then a threshold ϵ is utilized to construct a binary flag (S_{gene}) for gene selection employing the following equation:

$$S_{gene} = \begin{cases} 1, & \text{for } S_{gene} \geq \epsilon \\ 0, & \text{otherwise} \end{cases} \tag{4.21}$$

Each particle is a subset of informative genes, with each gene's efficacy in the subset represented by the number "1." The performance of the fitness function is then assessed using this informative gene subset as a candidate gene subset.

Increasing classification accuracy and choosing a manageable number of useful genes are the goals of this methodology. As a result, the fitness function in this example is built in a way that minimizes the classification error while also increasing classification accuracy and reducing the number of informative genes. The fitness function is so described as follows:

$$FF = E_{LOOCV} \times \omega_1 + \left(\frac{N_G}{M}\right) \times \omega_2 \tag{4.22}$$

where E_{LOOCV} is the error over leave-one-out-cross-validation accuracy (LOOCV), and it is evaluated as

$$E_{LOOCV} = 1 - A_{LOOCV} \tag{4.23}$$

where $A_{LOOCV} \in [0, 1]$ is the LOOCV accuracy. M represents the total number of genes used in the PSO-based wrapper method. N_G denotes the number of genes in the candidate gene subset. ω_1 and ω_2 are two priority weights corresponding to the error and the number of selected genes, respectively. The simplified diagram of this PSO-based gene identification technique is shown in Fig. 4.5, where, first, the gene expression value for each gene is shown. The representation of each particle in PSO for gene selection has been shown in Fig. 4.5 with a random value between "0" and "1." Then, the threshold ϵ has been used for each particle to have binary representation that actually helps to determine the selection of a particular gene. Finally, fitness function has been evaluated using Eq. 4.22.

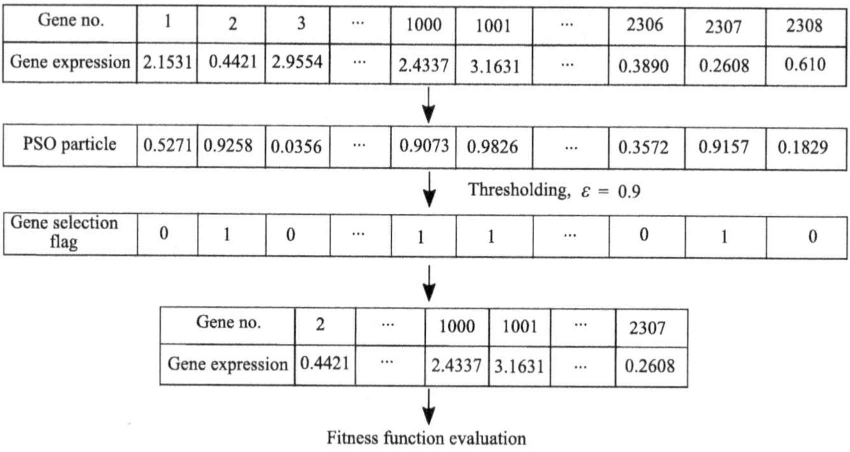

Fig. 4.5 Simplified diagram of PSO-based gene selection

4.4 Classifiers

Data input is mapped into a particular category of a general population using the mathematical technique of classification. Depending on how closely the data match, the population is split into a number of categories. These divisions are commonly referred to as classes. A classifier's job is to map the input data into a particular population class, so its purpose is to perform the classification task. We have a variety of classifiers at our disposal when it comes to machine learning. The following section provides an explanation of the two most well-known classifiers, support vector machine (SVM), and k-nearest neighbor (k-NN) classifiers.

4.4.1 *Support Vector Machine (SVM) Classifier*

SVM has been proposed by Vapnik [36]. It is a two-class classifier that divides the two data classes by building a hyperplane that maximizes the separation between the nearest data points of each class. All the data points that are located on one side of the hyperplane are members of the same class, and all the data points that are located on the opposite side are members of the opposite class. If x_i is the i_{th} input training sample, y_i is the corresponding target class, and n is the total number of training samples $\{(x_i, y_i), \forall i = 1, 2, \ldots, n\}$, the hyperplane can be defined as (vide Fig. 4.6)

$$(w.x_i) + b = 0 \tag{4.24}$$

where w is the weight vector, x is the training sample vector, and b is the bias. Each input training sample is connected to a weight vector, which conveys information about how relevant each input training sample is for differentiating between the two classes. A unique SVM parameter is bias. In essence, it is a value below which the hyperplane will always pass through the origin. There are two approaches to

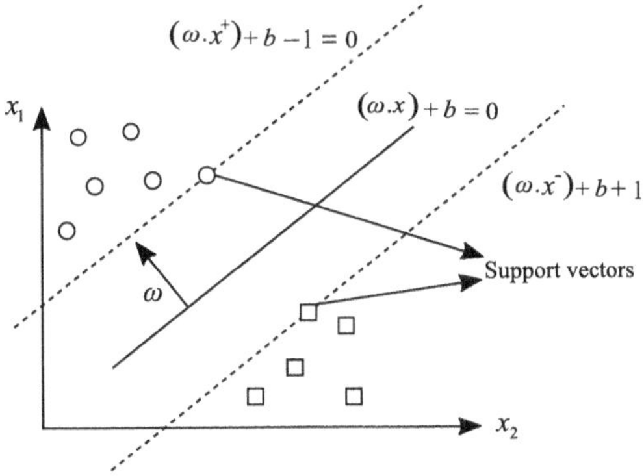

Fig. 4.6 Support vector machine

define the separating margin: functionally and geometrically. The functional margin can be defined from the hyperplane equation by multiplying the training vector x with the corresponding target class y, $y.((w.x) + b)$. The distance between the hyperplane and each data point determines the geometric margin. Critical points or support vectors are data points that are located very close to the hyperplane. Finding the support vectors is the purpose of SVM. The boundary of a class is defined by the support vectors. These support vectors contain all the information required for classifying new data. These support vectors of each class can define hyperplanes too with functional margin values +1 and −1. These hyperplanes can be defined as (vide Fig. 4.6)

$$\left(w.x^{+}\right) + b - 1 = 0 \tag{4.25}$$

$$\left(w.x^{-}\right) + b + 1 = 0 \tag{4.26}$$

where the sign over the x denotes the class. Therefore, the objective of SVM is to minimize the norm of the weight vector (w) such that the geometric margins of the data points maximize. Therefore, the mathematical formulation of the SVM has been given by

$$\text{minimize}_{w,b}\ \|w.w\| \quad \text{subject to} \quad y_i.\left((w.x_i) + b\right) \geq 1 \tag{4.27}$$

The aforementioned equation demonstrates how the geometric margin between two classes is maximized by minimizing the weight vector's norm. The problem has been turned into a compact formulation, where the objective and the constraint have been merged into one equation using Lagrangian multipliers, in order to simplify the previously mentioned mathematical formulation [36].

$$L\left(w, b, \alpha\right) = \frac{1}{2}\|w.w\| - \sum_{i=1}^{n} \alpha_i \left[y_i\left((w.x_i) + b\right) - 1\right] \tag{4.28}$$

where α are the Lagrange multipliers. Now, differentiating (4.28) with respect to w and b:

$$\frac{\partial L\left(w, b, \alpha\right)}{\partial w} \rightarrow w = \sum_{i=1}^{n} y_i \alpha_i x_i \tag{4.29}$$

$$\frac{\partial L\left(w, b, \alpha\right)}{\partial b} \rightarrow \sum_{i=1}^{n} y_i \alpha_i = 0 \tag{4.30}$$

Substituting Eq. (4.29) and Eq. (4.30) back into Eq. (4.28):

$$\text{maximize} \quad w(\alpha) = \sum_{i=1}^{n} \alpha_i - \frac{1}{2} \sum_{i=1}^{n} \sum_{j=1}^{n} y_i y_j \alpha_i \alpha_j \left(x_i . x_j\right)$$

$$\text{subject to} \quad \sum_{i=1}^{n} y_i \alpha_i = 0 \tag{4.31}$$

As a result, the training samples, which are often described as vectors, can be thought of as the linear combination that represents the equation's solution.

A linearly separable hyperplane is obtained using the aforementioned methods to linearly divide the two classes of data samples (vide Eq. 4.24). Since it is not always possible to produce a linearly separable hyperplane, the data are instead projected onto a high dimensional space from the original space, which aids in the creation of a linear hyperplane to help divide the two classes. Mathematically, a kernel function is denoted as $F\left(x_i, x_j\right) = \left(f\left(x_i\right), f\left(x_j\right)\right)$, where F is the kernel function and x_i, x_j are high dimensional input cases. Thus, the equation in (4.31) can be written using the kernel function in the following form:

$$w(\alpha) = \sum_{i=1}^{n} \alpha_i - \frac{1}{2} \sum_{i,j=1}^{n} y_i y_j \alpha_i \alpha_j F\left(x_i . x_j\right) \tag{4.32}$$

The kernel function $F\left(x_i . x_j\right)$ map the input data into a higher dimensional space where the data are linearly separable. The kernel function can be of different types such as linear kernel, polynomial kernel, quadratic kernel, and radial basic function (rbf) kernel [37].

Linear Kernel It is arguably the most basic kernel. It often has vectorial input data. The expression for the linear kernel function is

$$F\left(x_i . x_j\right) = x_i^T x_j \tag{4.33}$$

Polynomial Kernel Function The polynomial kernel function is expressed as

$$F\left(x_i . x_j\right) = \left(x_i^T x_j + 1\right)^d \tag{4.34}$$

where d is the order of polynomial kernel. The kernel has a tendency to overfit the training set if the order is set too high.

Quadratic Kernel A common variety of polynomial kernel is the quadratic kernel. The quadratic kernel function is written as

$$F\left(x_i . x_j\right) = \left(x_i^T x_j + c\right)^2 \tag{4.35}$$

where c is a constant that allows to trade off the influence of the higher-order and lower-order terms.

Rbf Kernel The rbf kernel is also known as Gaussian kernel. The rbf kernel is expressed as

$$F\left(x_i.x_j\right) = exp\left\{-\frac{\left\|x_i - x_j\right\|^2}{2\sigma^2}\right\} \tag{4.36}$$

where $\left|\left(x_i - x_j\right)\right|$ is the Euclidean distance between the data points. The distances are divided by σ and then normalized. The parameter controls the influence of each data point by acting as a smoothing parameter. The model cannot represent the complexity of data points if σ is set too low. On the other side, the kernel would be unable to stop overfitting to the training data if the value of σ is too high.

The best method for creating a K class SVM, however, is to develop a number of binary classifiers and combine them, where each binary classifier is trained to distinguish one class from the other $K - 1$ classes.

4.4.2 *k-Nearest Neighbor Classifier*

A reliable and benchmark classifier is the k-NN [38]. Additionally, it is non-parametric, which essentially means that it makes no assumptions about the distribution of the data [39]. It is pretty straightforward conceptually and simple to implement. If n is the total number of training samples with k number of classes, $\{(x_1, y_1), (x_2, y_2), \ldots, (x_n, y_k)\}$, where x_i represents the vector of training samples and y_i represents the corresponding classes, then for a given test sample x_j, its distance to each labeled sample has been measured. For k-NN classifier, the test sample x_j is assigned to the class of its closest labeled sample if the number of neighborhood (k) is equal to 1. The label y_j of the test sample x_j can then be expressed as

$$y_j = min_{\forall x_j}\left\{D_{x_i}\left(x_j\right)\right\} \tag{4.37}$$

where $D_{x_i}\left(x_j\right)$ is the distance between the test sample x_j and the training sample x_i. Still some redundant attributes may remain present, and they cause degradation of classification performance. This is, more popularly, known as "curse of dimensionality."

The test's sample level y_j is determined by the majority votes of the neighborhood classes if the value of neighborhood k is set to be greater than 1. The classifier becomes robust and less sensitive to noise as a result of assuming a neighborhood value greater than 1 [39]. A appropriate neighborhood value must be chosen to improve the performance of the classifier.

4.5 Experimental Study

Three benchmark microarray gene expression data sets namely ALL_AML data, MLL data, and SRBCT data are used in this chapter to study the effectiveness of filter-based and wrapper-based gene selection techniques. These data sets contain two, three, and four classes, respectively.

4.5.1 Experimental Study on Filter Approaches

In the filter method-based gene selection technique, all the genes are ranked according to the three statistical gene ranking methods, namely, t-test, chi-squared statistics, and signal-to-noise ratio. Top 100 genes are picked up with the highest statistical scores to perform the classification task. First, the gene with the highest score is put to the classifiers. This process is repeated with top two genes and then top three genes and so on. SVM classifier with different kernels and *k*-NN classifier are used for classification purpose in this chapter [11, 40].

Three statistical filter methods namely t-test, chi-squared statistics, and SNR have been applied on ALL_AML, MLL, and SRBCT data to rank all the genes according to their importance. Then the top-ranked gene has been fed to the classifiers, and the corresponding blind classification accuracies are evaluated. Then the same process has been repeated with two top-ranked genes, then with three top-ranked genes, and so on. The blind test accuracies of top 100 genes for all the three benchmark data sets with different SVM kernels and *k*-NN classifier are shown in Figs. 4.7, 4.8, and 4.9. In the figures, the red, blue, pink, black, and green curves show the classification accuracies of top 100 genes, when fed one by one into the different classifiers, namely *k*-NN, SVM (linear), SVM (polynomial), SVM (quadratic), and SVM (rbf), respectively. The best accuracies achieved using different filter method-based gene selection techniques for the three data sets are shown in Tables 4.1, 4.2, and 4.3, respectively.

A test accuracy of 100% is achieved in SRBCT data with top 48 genes while using the combination of chi-squares statistics and *k*-NN classifier. The scatter plots of 48 genes have been shown in Figs. 4.10, 4.11, 4.12, 4.13, 4.14, and 4.15. In these figures, each panel corresponds to one gene. The red, green, blue, and black circles correspond to EWS, NHL, NB, and RMS classes, respectively. The EWS class is moderate to highly expressed for the panel numbers 1, 3, 7, 10, 12, 14, 16, 17, 18, 20, 35, 37, 39, 40, 43, 44, 45, 46. The NHL class is moderate to highly expressed for panel numbers 2, 6, 8, 12, 14, 19, 24, 25, 28, 32, 34, 41, 42. The NB class is moderate to highly expressed for panel numbers 2, 4, 6, 7, 8, 9, 11, 12, 18, 19, 22, 23, 25, 28, 31, 34, 41, 42, and the RMS class is moderate to highly expressed for panel numbers 6, 7, 9, 11, 14, 19, 23, 25, 26, 33, 44, 45.

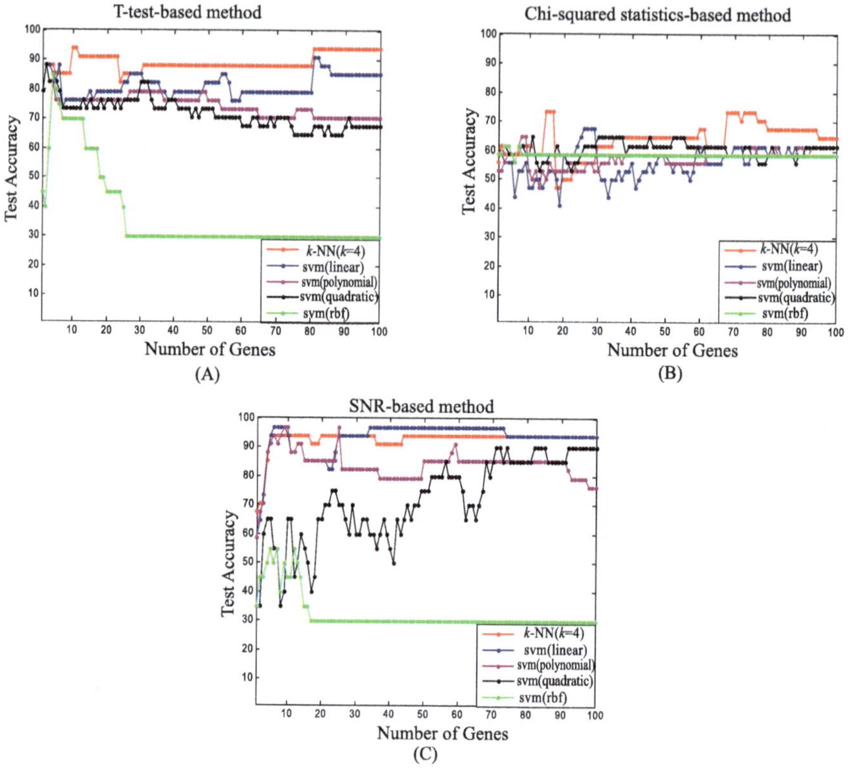

Fig. 4.7 Test accuracy of top 100 genes for ALL_AML data. (**a**) T-test-based method. (**b**) Chi-squared statistics-based method. (**c**) SNR-based method

A test accuracy of 97.0588% is achieved in ALL_AML data with top 6 genes while using signal-to-noise ratio and SVM classifier with linear kernel. The scatter plots of 6 genes have been shown in Fig. 4.16 where each panel corresponds to one gene. The red and blue colors correspond to ALL and AML classes, respectively. The scatter plot shows that the ALL class is moderately expressed in genes M30703_s_at, D88422_at and M31166_at. The AML class is moderate to highly expressed in genes J04990_at, D28235_s_at, D88422_at and M84526_at.

A test accuracy of 100% is achieved with MLL data using top 2 genes with t-test method and SVM classifier with linear kernel. The scatter plots of 2 genes have also been shown in Fig. 4.17 where each panel corresponds to one gene. The red, blue, and green colors correspond to ALL, AML, and MLL classes, respectively. The scatter plot shows that the ALL class is highly expressed in genes 266_s_at and 1389_at. The AML and MLL classes are moderate to highly expressed in gene 266_s_at.

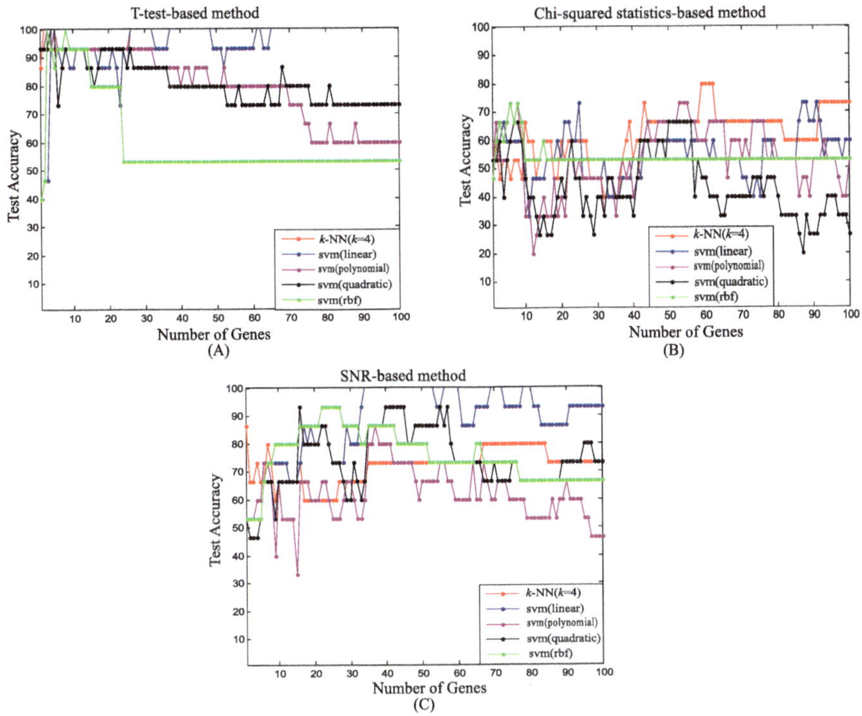

Fig. 4.8 Test accuracy of top 100 genes for MLL data. (**a**) T-test-based method. (**b**) Chi-squared statistics-based method. (**c**) SNR-based method

The pseudo-color images of the ALL_AML, MLL, and SRBCT data sets have been shown in Figs. 4.18, 4.19, and 4.20, respectively. The pseudo-color images of the ALL_AML, MLL, and SRBCT data sets have been shown in Figs. 4.18, 4.19, and 4.20, respectively. Each panel represents a specific class of the data sets in the pseudo-color images. The change in color from blue to red represents the alteration in the level of expression of a particular gene chosen for that class.

The list of best subsets of selected genes by using different filter methods have also been shown in Tables 4.4, 4.5, and 4.6, respectively.

4.5.2 *Experimental Study on Wrapper Approaches*

In the case of PSO-based wrapper method, first, t-test technique is applied to select 500 top-ranked genes from the three benchmark data sets. In order to actively participate in the gene selection process, these top 500 genes are subsequently used in the PSO-based wrapper technique. Each data set has been subjected to ten experiments using this wrapper technique, and the average outcomes are assessed. Being

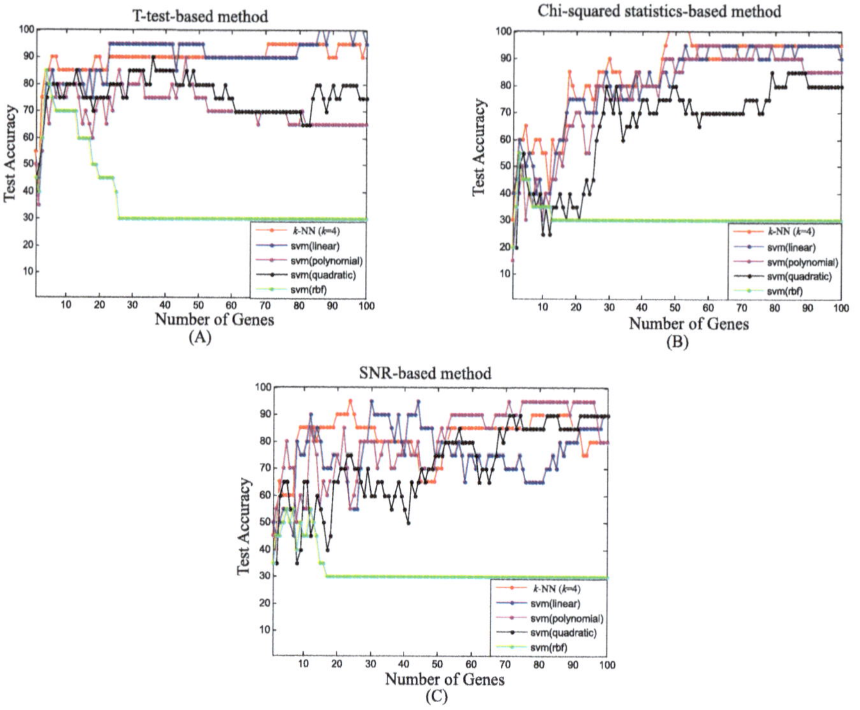

Fig. 4.9 Test accuracy of top 100 genes for SRBCT data. (**a**) T-test-based method. (**b**) Chi-squared statistics-based method. (**c**) SNR-based method

a stochastic approach, the PSO-based wrapper technique's calculation of average results is essential. It is important to note that fewer decisive genes are needed and high blind test accuracies are both necessary for exemplary performance. The results of ten independent runs on the three benchmark data sets are shown in Tables 4.7, 4.8, and 4.9 where the best results are shown in the shaded cells. The list of best subsets of selected genes by using PSO-based wrapper method have also been shown in Tables 4.10, 4.11, and 4.12, respectively.

Table 4.1 Test accuracy of ALL-AML data using different filter-based methods

Data	ALL_AML														
Method	T-test					Chi-squared statistics					SNR				
Classifier	*k*-NN	SVM linear	SVM polynomial	SVM quadratic	SVM rbf	*k*-NN	SVM linear	SVM polynomial	SVM quadratic	SVM rbf	*k*-NN	SVM linear	SVM polynomial	SVM quadratic	SVM rbf
No. of gene	10	81	2	2	4	15	26	8	11	3	5	6	9	71	5
Test accuracy	94.12	91.18	88.24	88.24	85	73.53	67.65	64.71	64.71	61.76	94.12	97.06	97.06	90	55

Table 4.2 Test accuracy of MLL data using different filter-based methods

Data	MLL														
Method	T-test					Chi-squared statistics					SNR				
Classifier	*k*-NN	SVM linear	SVM polynomial	SVM quadratic	SVM rbf	*k*-NN	SVM linear	SVM polynomial	SVM quadratic	SVM rbf	*k*-NN	SVM linear	SVM polynomial	SVM quadratic	SVM rbf
No. of gene	2	4	3	4	3	59	25	8	7	6	1	34	37	16	22
Test accuracy	100	100	100	100	100	80	73.33	73.33	66.66	73.33	86.66	100	86.66	93.33	93.33

Table 4.3 Test accuracy of SRBCT data using different filter-based methods

Data	SRBCT														
Method	T-test					Chi-squared statistics					SNR				
Classifier	*k*-NN	SVM linear	SVM polynomial	SVM quadratic	SVM rbf	*k*-NN	SVM linear	SVM polynomial	SVM quadratic	SVM rbf	*k*-NN	SVM linear	SVM polynomial	SVM quadratic	SVM rbf
No. of gene	71	86	46	36	4	48	53	57	79	3	24	30	71	71	5
Test accuracy	95	100	90	90	85	100	95	95	85	55	95	95	95	90	55

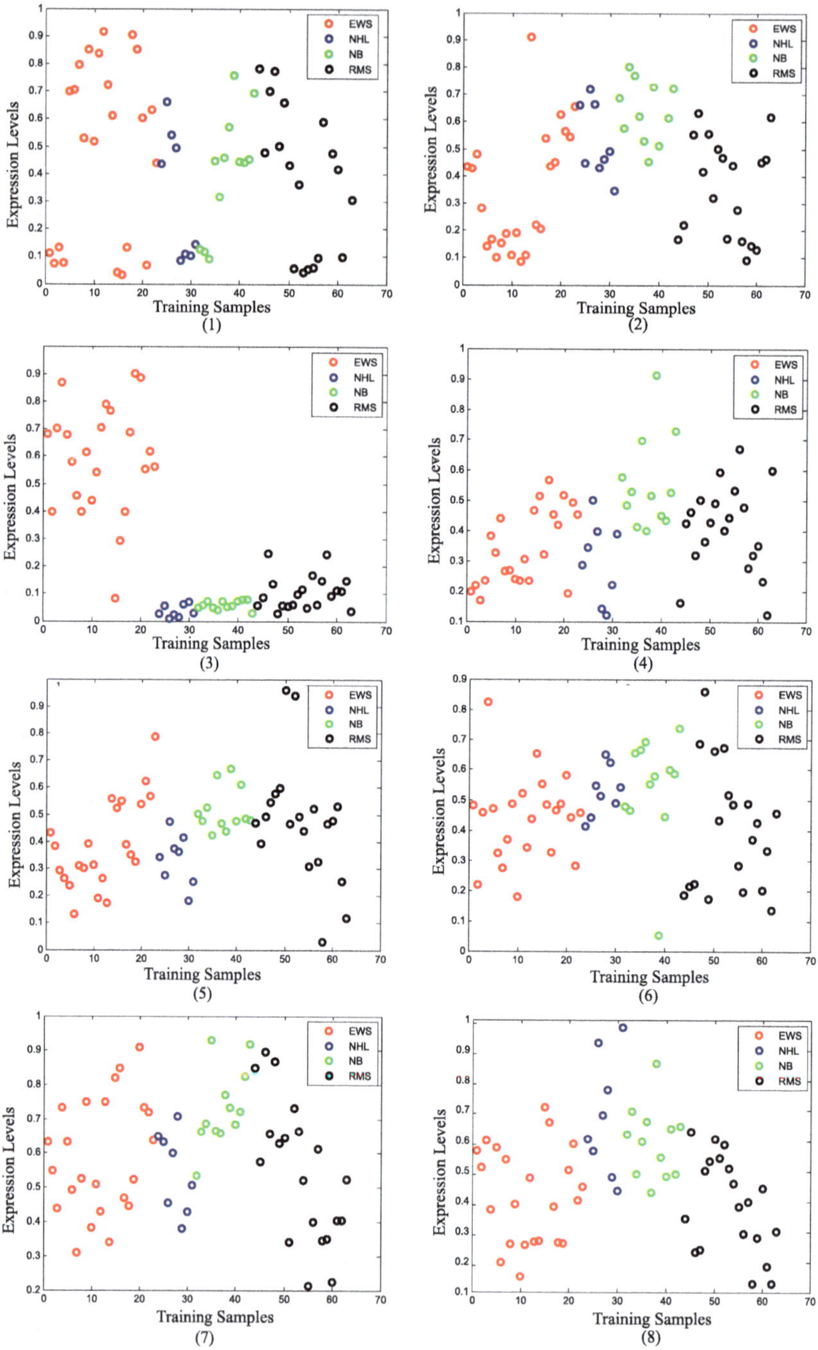

Fig. 4.10 Scatter plot of 1 to 8 genes from the top 48 genes of SRBCT data using filter method

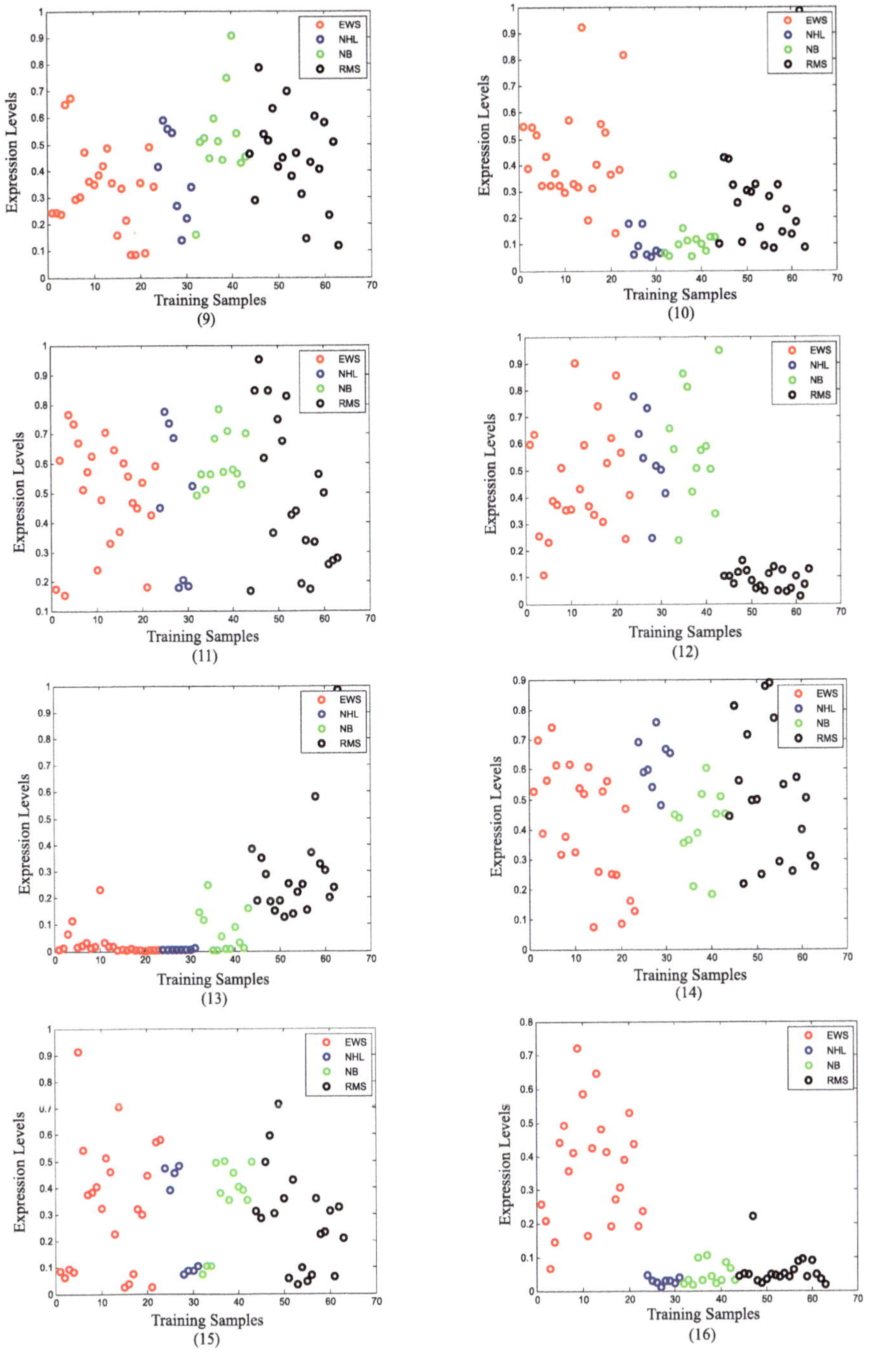

Fig. 4.11 Scatter plot of 9 to 16 genes from the top 48 genes of SRBCT data using filter method

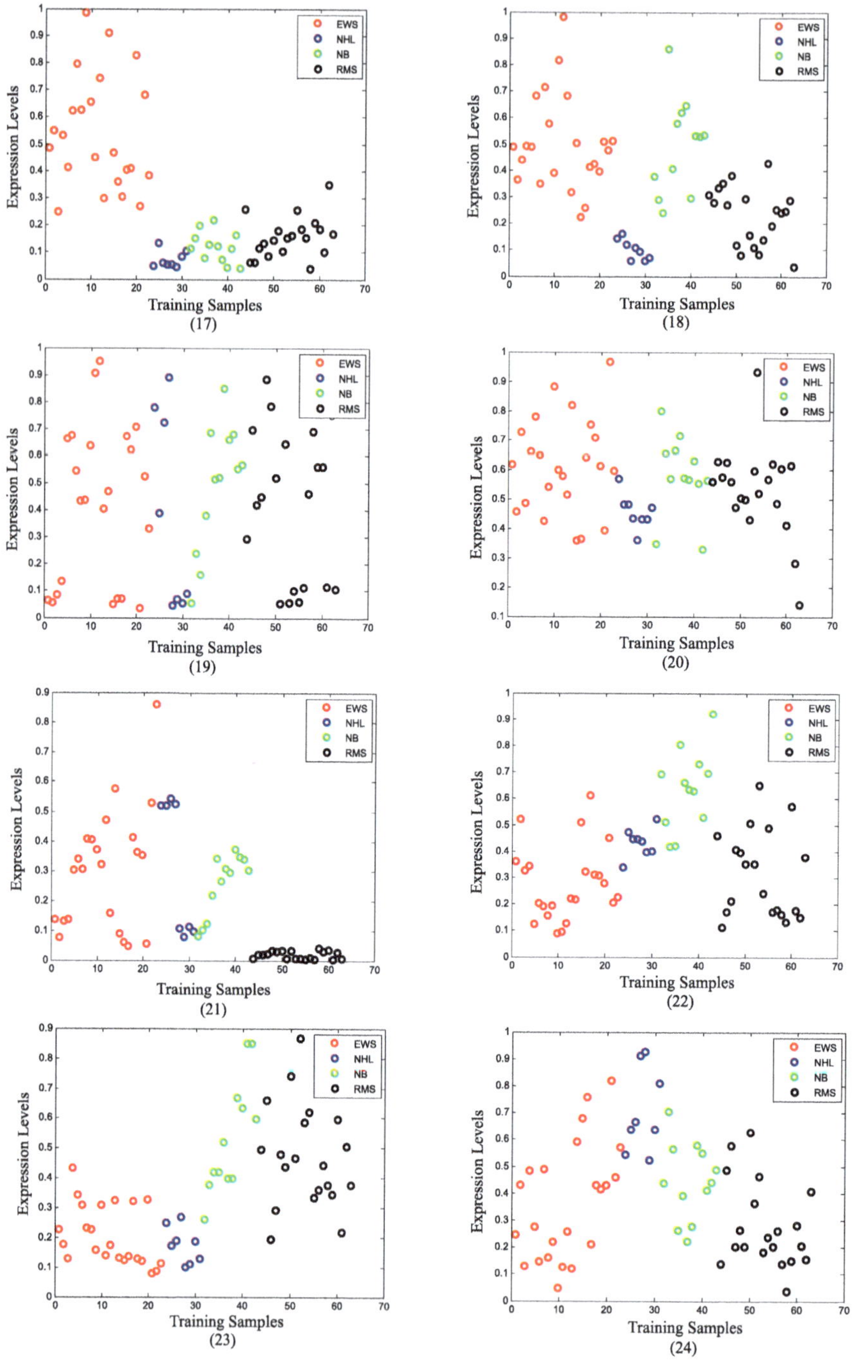

Fig. 4.12 Scatter plot of 17 to 24 genes from the top 48 genes of SRBCT data using filter method

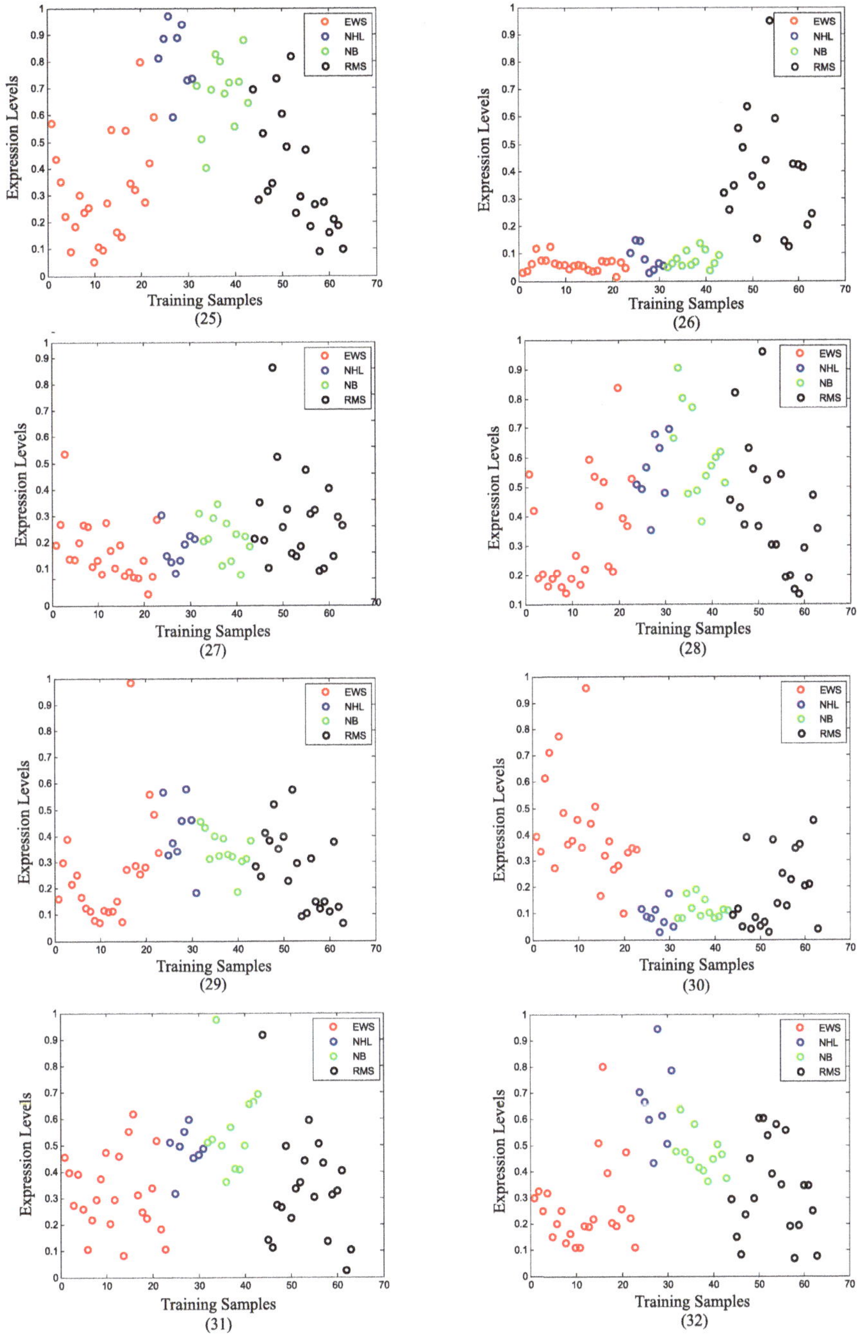

Fig. 4.13 Scatter plot of 25 to 32 genes from the top 48 genes of SRBCT data using filter method

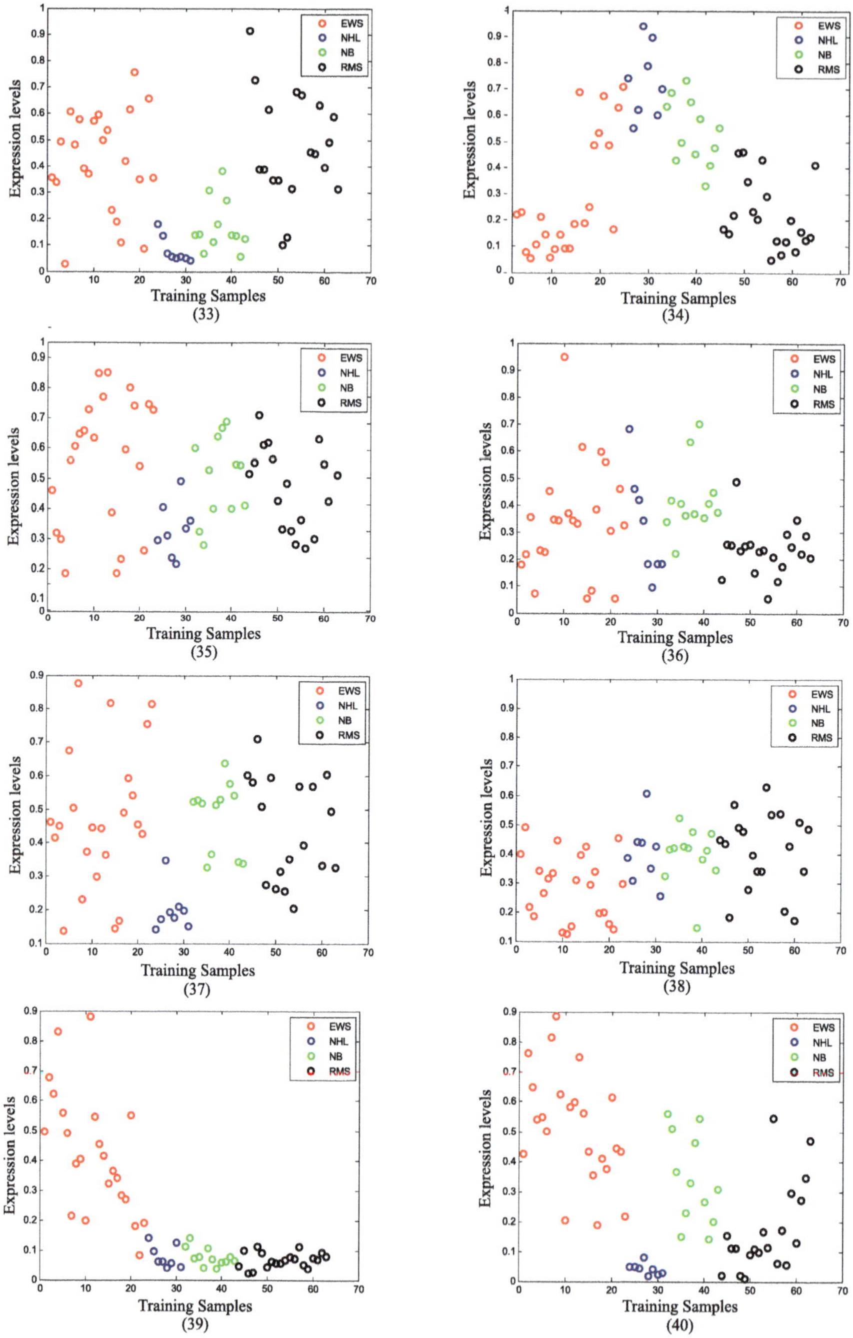

Fig. 4.14 Scatter plot of 33 to 40 genes from the top 48 genes of SRBCT data using filter method

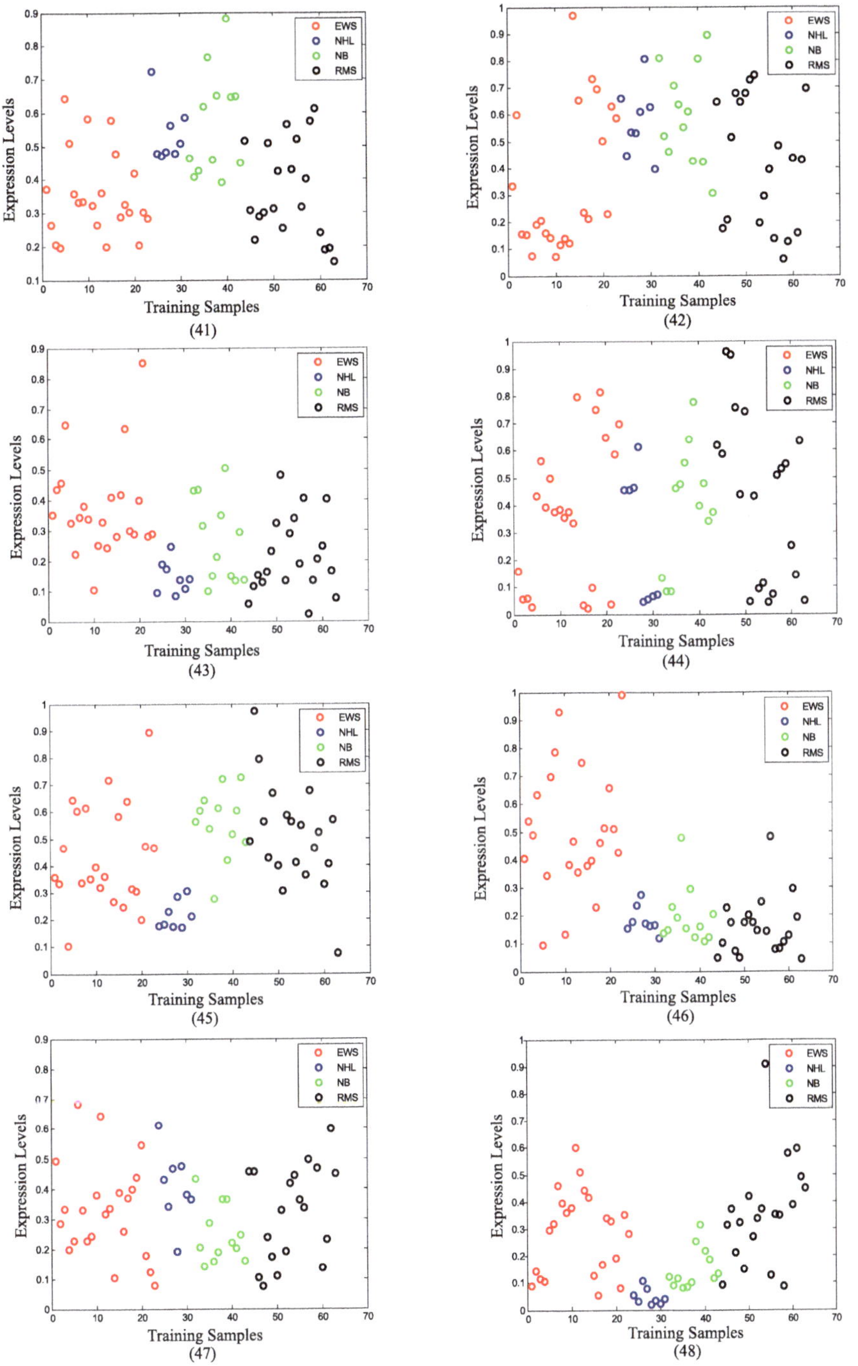

Fig. 4.15 Scatter plot of 41 to 48 genes from the top 48 genes of SRBCT data using filter method

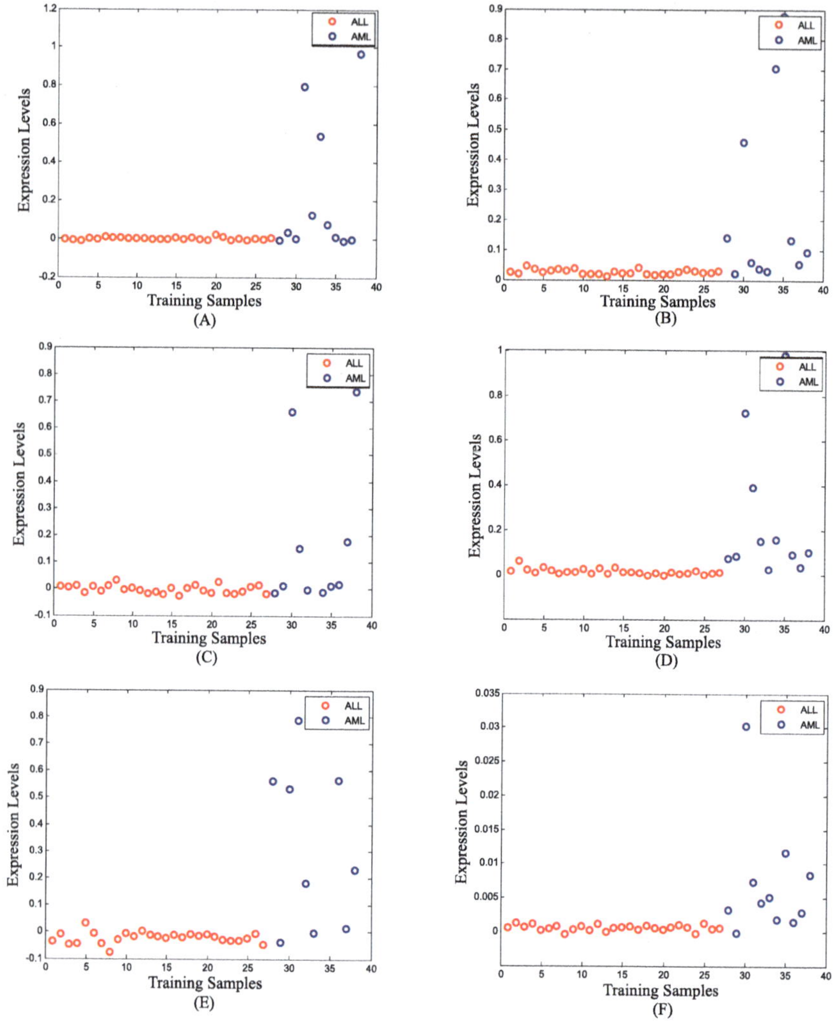

Fig. 4.16 Scatter plot of top 6 genes of ALL_AML data using filter method. (**a**) M30703_s_at, (**b**) J04990_at, (**c**) D28235_s_at, (**d**) D88422_at, (**e**) M84526_at, (**f**) M31166_at

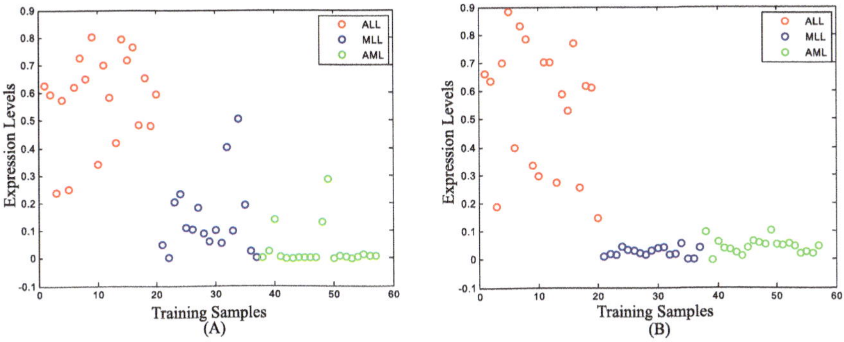

Fig. 4.17 Scatter plot of top 2 genes of MLL data using filter method. (**a**) 266_s_at, (**b**) 1389_at

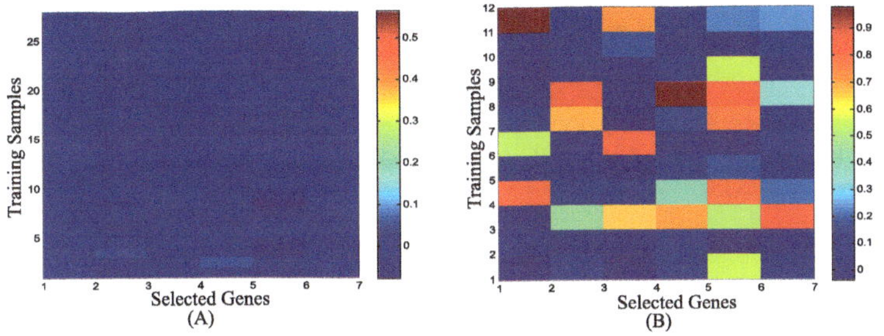

Fig. 4.18 Pseudo-color image of top 6 genes of ALL_AML data using filter method. (**a**) ALL. (**b**) AML

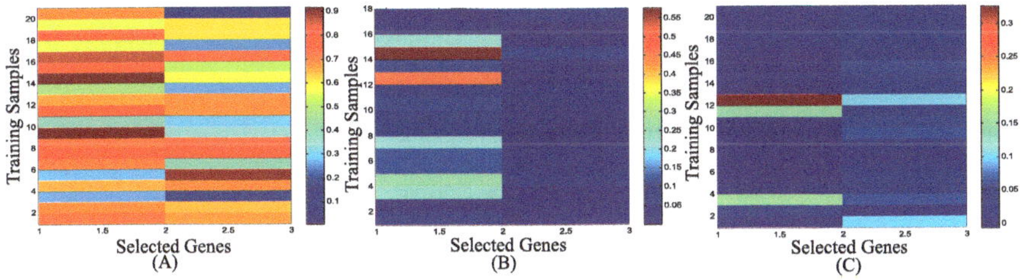

Fig. 4.19 Pseudo-color image of top 2 genes of MLL data using filter method. (**a**) ALL. (**b**) AML. (**c**) MLL

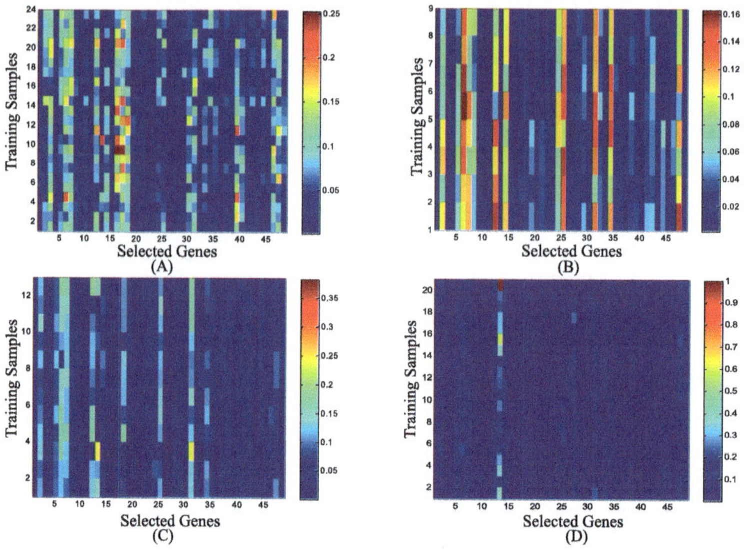

Fig. 4.20 Pseudo-color image of top 48 genes of SRBCT data using filter method. (**a**) EWS. (**b**) NHL. (**c**) NB. (**d**) RMS

Table 4.4 List of 6 genes for ALL_AML data using filter method

Index no.	Gene accession number	Gene description
6277	M30703_s_at	Amphiregulin (AR) gene
1133	J04990_at	CATHEPSIN G PRECURSOR
5599	D28235_s_at	Cyclooxygenase-2 (hCox-2) gene
760	D88422_at	CYSTATIN A
2288	M84526_at	DF D component of complement (adipsin)
1926	M31166_at	PTX3 Pentaxin-related gene, rapidly induced by IL-1 beta

Table 4.5 List of 2 genes for MLL data using filter method

Index no.	Gene accession number	Gene description
12418	266_s_at	"Hs.278667 gnl|*UG*|Hs#S885 Homo sapiens CD24 signal transducer mRNA, complete cds and 3' region"
11297	1389_at	"Hs.1298 gnl|*UG*|Hs#S1945 Human common acute lymphoblastic leukemia antigen (CALLA) mRNA, complete cds"

Relation of error, the number of genes, and fitness function with the progress of iteration for ALL_AML, MLL, and SRBCT data sets are shown in Figs. 4.21, 4.22, and 4.23, respectively. The trend of error is declining as iteration progresses, and the trend of gene numbers is likewise declining. Given that the fitness function

Table 4.6 List of 48 genes for SRBCT data using filter method

Index no.	Image ID	Gene description
282	856650	ATP synthase, H+ transporting, mitochondrial F1 complex, delta subunit
912	824041	Splicing factor, arginine/serine-rich
1389	770394	Fc fragment of IgG, receptor, transporter, alpha
453	487296	Matrix metalloproteinase 11 (stromelysin 3)
353	1455641	Proteasome (prosome, macropain) subunit, alpha type, 7
886	897567	Lactate dehydrogenase A
1632	884822	Zinc finger protein 6 (CMPX1)
1686	564756	Ras-GTPase-activating protein SH3-domain-binding protein
1432	855684	Golgi autoantigen, golgin subfamily a, 3
607	811108	Thyroid hormone receptor interactor 6
648	246549	Proteasome (prosome, macropain) subunit, beta type, 2
1412	266106	Tyrosine 3-monooxygenase/tryptophan 5-monooxygenase activation protein, epsilon polypeptide
509	207274	Human DNA for insulin-like growth factor II (IGF-2); exon 7 and additional ORF
1543	815303	Aspartyl-tRNA synthetase
1170	51814	Cystatin B (stefin B)
246	377461	Caveolin 1, caveolae protein, 22kD
545	1435862	Antigen identified by monoclonal antibodies 12E7, F21, and O13
951	841620	Dihydropyrimidinase-like 2
274	384078	–
1316	877782	Ras-related C3 botulinum toxin substrate 1 (rho family, small GTP binding protein Rac1)
1819	241365	Mannan-binding lectin serine protease 1 (C4/C2 activating component of Ra-reactive factor)
2258	810864	ESTs, Highly similar to CGI-48 protein [H.sapiens]
1888	774502	Protein tyrosine phosphatase, non-receptor type 12
76	725454	CDC28 protein kinase 2
937	789204	Translocation protein 1
1003	796258	Sarcoglycan, alpha (50kD dystrophin-associated glycoprotein)
1955	784224	Fibroblast growth factor receptor 4
450	451907	–
1685	758343	ESTs
1772	713922	Glutathione S-transferase M1
25	34355	Calmodulin 2 (phosphorylase kinase, delta)
2276	502421	ESTs, Highly similar to COBW-like placental protein [H.sapiens]
719	159455	Similar to vaccinia virus HindIII K4L ORF
1932	782811	High-mobility group (nonhistone chromosomal) protein isoforms I and Y
807	770216	Ankyrin repeat-containing protein
939	713782	A disintegrin and metalloproteinase domain 15 (metargidin)
239	809720	Centromere protein B (80kD)

(continued)

Table 4.6 (continued)

Index no.	Image ID	Gene description
702	196501	Arginine-rich protein
1954	814260	Follicular lymphoma variant translocation 1
1980	841641	Cyclin D1 (PRAD1: parathyroid adenomatosis 1)
1784	665774	Eukaryotic translation initiation factor 4E
1025	564803	Forkhead (Drosophila)-like 16
891	75923	Human zinc finger protein mRNA, complete cds
576	49591	Steroid sulfatase (microsomal), arylsulfatase C, isozyme S
841	809916	Zinc finger protein 144 (Mel-18)
1074	1471841	ATPase, Na+/K+ transporting, alpha 1 polypeptide
475	1472753	Eukaryotic translation initiation factor 4B
1795	45291	Dentatorubral-pallidoluysian atrophy (atrophin-1)

is created by adding the gene counts and CV accuracy, its tendency is also on the decline as iterations advance. This pattern implies that a limited selection of interesting genes is discovered by the PSO-based wrapper technique, which can improve classification performance.

The scatter plots of the selected genes for the ALL_AML data set have been shown in Fig. 4.24. In the scatter plot, each panel corresponds to one gene. The figure shows that the ALL class is moderate to highly expressed in genes AFFX-BioDn-5_st and D17716_at. The AML class is moderate to highly expressed in genes AFFX-BioDn-5_st, D14664_at and D17716_at. The scatter plots of the selected genes for the MLL data set have been shown in Fig. 4.25. In the scatter plot, each panel corresponds to one gene. The figure shows that the ALL class is moderate to highly expressed in genes 31351_at, 31412_at, 31553_at, 31704_at. The AML class is moderate to highly expressed in genes 31351_at, 31412_at, 31553_at, 31686_at, 31704_at, and the MLL class is moderate to highly expressed in genes 31351_at, 31353_f_at, 31473_s_at, 31553_at, 31686_at, 31704_at. The scatter plots of the selected genes for the SRBCT data set have been shown in Figs. 4.26 and 4.27. In the scatter plots, each panel corresponds to one gene. The figure shows that the EWS class is moderate to highly expressed in genes termed as Image ID 814260, Image ID 491565, Image ID 789204, Image ID 950367, Image ID 75923. The NHL class is moderate to highly expressed in genes termed as Image ID 789204, Image ID 781047, Image ID 51041, Image ID 950367. The NB class is moderate to highly expressed in genes termed as Image ID 812105, Image ID 377048, Image ID 789204, Image ID 51041, Image ID 364510, and the RMS class is moderate to highly expressed for genes termed as Image ID 839552, Image ID 789204, Image ID 813841, Image ID 1409509, Image ID 753620.

Table 4.7 Result of ten independent runs on ALL_AML data set using PSO-based wrapper method

Run	No. of genes	Index no. of selected genes	Error	Training acc.	Test acc.
1	8	33,81,108,111,159,202,314,436	0	100	85.29
2	11	20,27,47,62,81,142,189,206,230,256,498	0	100	82.35
3	11	76,81,121,202,224,230,401,428,452,461,498	0	100	85.29
4	11	3,51,65,69,71,106,114,186,314,315,357	0	100	61.76
5	13	4,27,46,76,108,171,179,189,204,314,321,395,453	0	100	85.29
6	10	12,27,39,81,142,230,251,343,357,498	0	100	88.23
7	15	27,51,75,76,127,131,159,204,256,314,321,349,420,453,462	0	100	88.35
8	6	130,144,157,357,420,484	0	100	91.18
9	3	15,235,274	0	100	91.18
10	9	62,121,142,156,249,255,363,427,498	0	100	61.76
Avg.	9.70	–	0	100	81.46
Std.	3.43	–	0	0	10.83

Table 4.8 Result of ten independent runs on MLL data set using PSO-based wrapper method

Run	No. of genes	Index no. of selected genes	Error	Training acc.	Test acc.
1	6	16,106,138,155,494,497	0	100	73.33
2	8	3,77,105,216,290,312,429,494	0	100	86.66
3	7	94,96,155,216,296,429,447	0	100	100
4	4	2,96,216,296	0	100	86.66
5	5	16,263,303,416,429	0	100	73.33
6	4	16,32,40,283	0	100	80
7	4	32,40,183,472	0	100	80
8	4	6,32,274,283	0	100	80
9	6	32,109,130,303,374,472	0	100	86.66
10	5	40,96,230,246,444	0	100	66.67
Avg.	5.30	–	0	100	81.33
Std.	1.41	–	0	0	9.32

Table 4.9 Result of ten independent runs on SRBCT data set using PSO-based wrapper method

Run	No. of genes	Index no. of selected genes	Error	Training acc.	Test acc.
1	12	589,1954,1327,174,910,970,1708,372,1924,1372,1656,796	0	100	70
2	14	1601,1327,276,1084,1723,1645,368,1708,1142,142,554,137,1298,373	0	100	85
3	7	1955,1327,1084,1723,1708,1656,1580	0	100	80
4	10	783,545,2144,2046,1084,1723,137,1469,662,643	0	100	75
5	9	742,1601,1327,1723,1645,142,1580,643,1867	0	100	90
6	12	1955,1327,276,975,937,1723,368,1708,142,1091,2240,1962	0	100	70
7	10	545,1453,2046,1207,975,424,1924,142,230,643	0	100	80
8	7	1955,1319,2276,1708,142,2022,1201	0	100	90
9	7	742,910,1319,2253,1708,1924,657	0	100	85
10	14	742,1954,1662,1327,910,937,1723,1414,1055,1887,142,1962,891,1929	0	100	100
Avg.	10.20	–	0	100	82.50
Std.	1.41	–	0	0	9.46

Table 4.10 List of 3 genes for ALL_AML data using PSO-based wrapper method

Index no.	Gene accession number	Gene description
15	AFFX-BioDn-5_st	AFFX-BioDn-5_st (endogenous control)
235	D14664_at	KIAA0022 gene
274	D17716_at	GB DEF = N-acetylglucosaminyltransferase V

Table 4.11 List of 7 genes for MLL data using PSO-based wrapper method

Index no.	Gene accession number	Gene description
94	31351_at	Hs.135631 gnl|*UG*|Hs#S553186 H.sapiens synthetic gene for platelet-derived growth factor-BB
96	31353_f_at	Hs.135634 gnl|*UG*|Hs#S572250 H.sapiens HFKH4 mRNA for fork head-like protein
155	31412_at	Hs.158343 gnl|*UG*|Hs#S891817 Homo sapiens testis-specific PTP-BL-Related Y protein (PRY) mRNA, complete cds
216	31473_s_at	Hs.131814 gnl|*UG*|Hs#S1263995 Homo sapiens truncated TRF1-interacting ankyrin-related ADP-ribose polymerase TT7 mRNA, alternatively spliced, partial cds
296	31553_at	Hs.99971 gnl|*UG*|Hs#S4302 H.sapiens HZF8 mRNA for zinc finger protein
429	31686_at	Hs.234761 gnl|*UG*|Hs#S592083 Human beta-tubulin (TUB4q) gene, complete cds
447	31704_at	Hs.103503 gnl|*UG*|Hs#S472758 Human DNase 1 homolog (DNAS1L2) mRNA, complete cds

Table 4.12 List of 14 genes for SRBCT data using PSO-based wrapper method

Index no.	Image ID	Gene description
742	812105	Transmembrane protein
1954	814260	Follicular lymphoma variant translocation 1
1662	377048	Homo sapiens incomplete cDNA for a mutated allele of a myosin class I, myh-1c
1327	491565	Cbp/p300-interacting transactivator, with Glu/Asp-rich carboxy-terminal domain, 2
910	839552	Nuclear receptor coactivator 1
937	789204	Translocation protein 1
1723	813841	Plasminogen activator, tissue
1414	781047	Budding uninhibited by benzimidazoles 1 (yeast homolog)
1055	1409509	Troponin T1, skeletal, slow
1887	51041	ESTs
142	364510	Special AT-rich sequence binding protein 1 (binds to nuclear matrix/scaffold-associating DNAs)
1962	950367	Adenosine deaminase, RNA-specific
891	75923	Human zinc finger protein mRNA, complete cds
1929	753620	Insulin-like growth factor binding protein 6

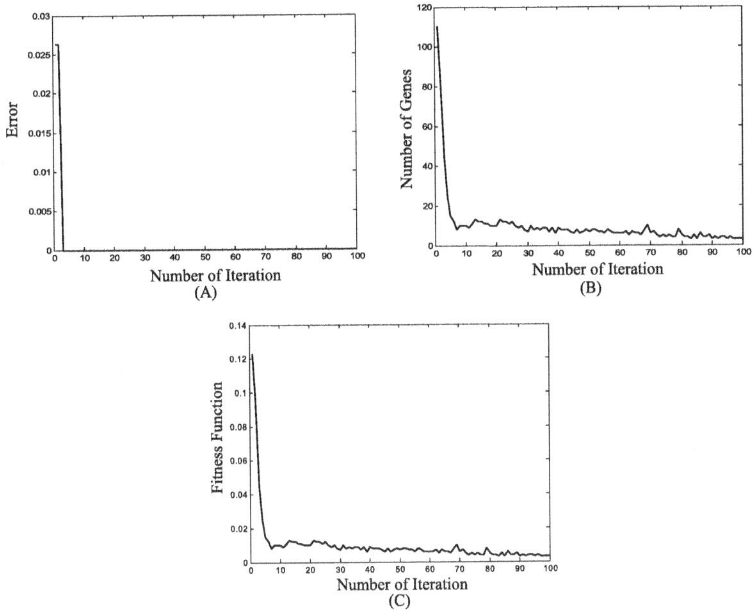

Fig. 4.21 Result after 100 PSO iterations on ALL_AML data. (**a**) Error is decreasing with the number of iterations, (**b**) the number of genes is decreasing with the number of iterations, (**c**) fitness function is decreasing with the number of iterations

The pseudo-color images of the best subsets of ALL_AML, MLL, and SRBCT data sets using PSO-based wrapper method have been shown in Figs. 4.28, 4.29, and 4.30, respectively. In the pseudo-color images, each panel represents one particular class of the data sets. The transition of color from blue to red demonstrates the change of expression levels of a particular selected gene for that particular class.

4.6 Summary

The gene selection challenge, which entails finding the pertinent or helpful genes in microarray gene expression data, has been investigated in this chapter. The t-test, chi-squared statistics, and SNR are a few approaches with filter-based methodologies that have been contrasted. The top-ranked genes from filter-based approaches, however, may not always correspond to the important genes. The top-ranked genes may include some redundant genes. In order to isolate the tiny group of meaningful genes from the top 500 genes, the PSO-based wrapper technique is employed in this chapter. The PSO-based wrapper technique can successfully identify small subsets of important genes for precise cancer subgroup prediction.

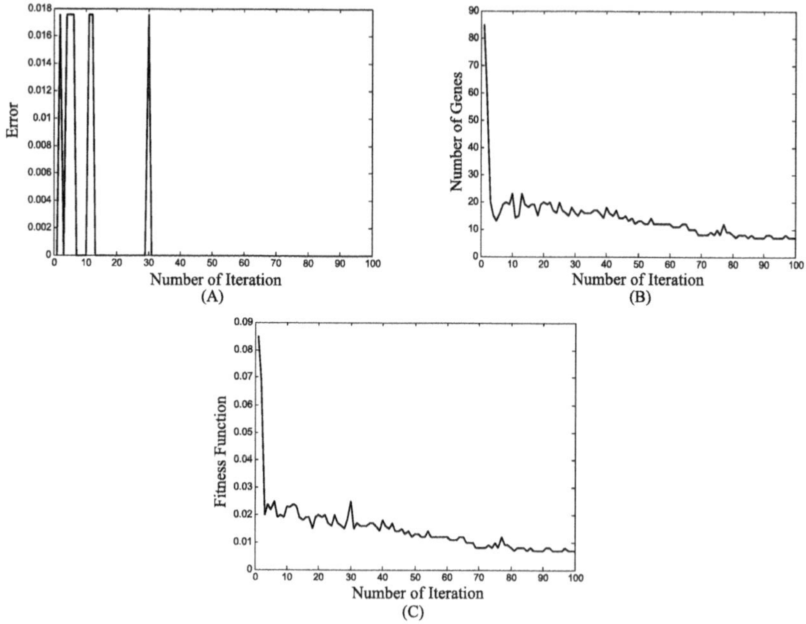

Fig. 4.22 Result after 100 PSO iterations on MLL data. (**a**) Error is decreasing with the number of iterations, (**b**) the number of genes is decreasing with the number of iterations, (**c**) fitness function is decreasing with the number of iterations

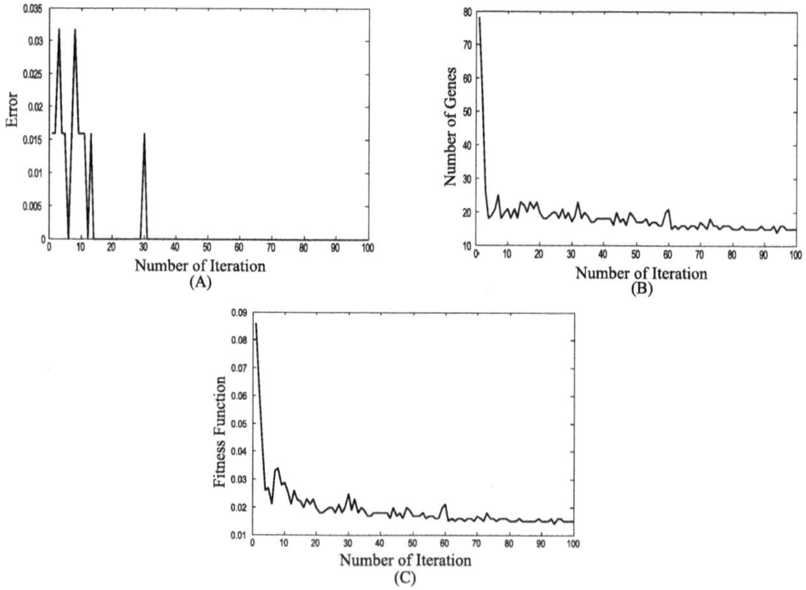

Fig. 4.23 Result after 100 PSO iterations on SRBCT data. (**a**) Error is decreasing with the number of iterations, (**b**) the number of genes is decreasing with the number of iterations, (**c**) fitness function is decreasing with the number of iterations

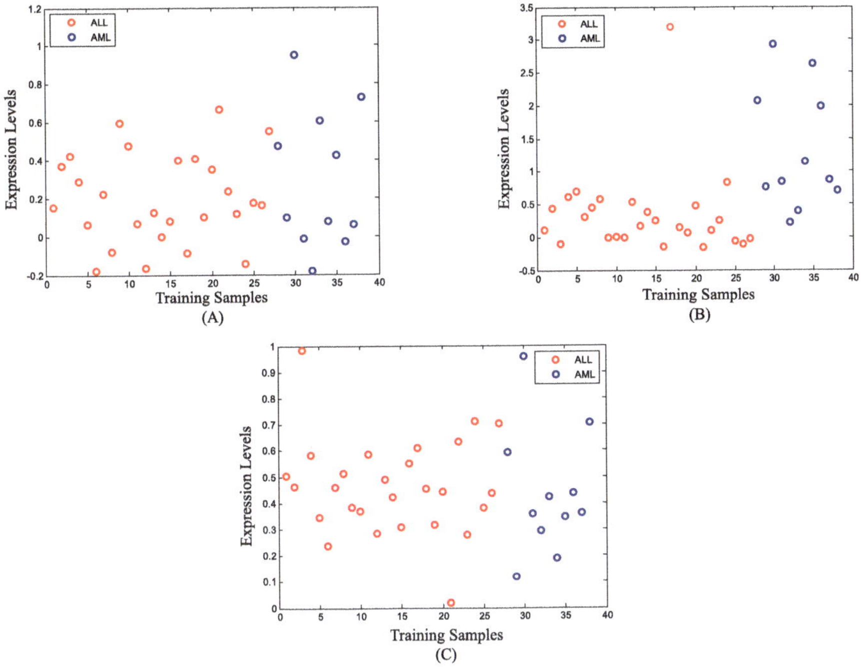

Fig. 4.24 Scatter plot of selected 3 genes of ALL_AML data using PSO-based wrapper method. (**a**) AFFX-BioDn-5_st, (**b**) D14664_at, (**c**) D17716_at

However, in this chapter, statistical gene ranking method like the t-test is used to reduce the dimensionality of the data sets to 500 genes. Then, for this smaller set of genes, only the PSO-based wrapper approach is used. As a result, a larger number of genes are completely ignored when using the PSO-based wrapper method, while a smaller number of genes are merely taken into account. These numerous genes also contain several valuable genes that can be used to precisely classify the various types of cancer. The component of gene selection based on a wrapper-based strategy that clearly outperforms a filter-based approach is highlighted in the following chapter.

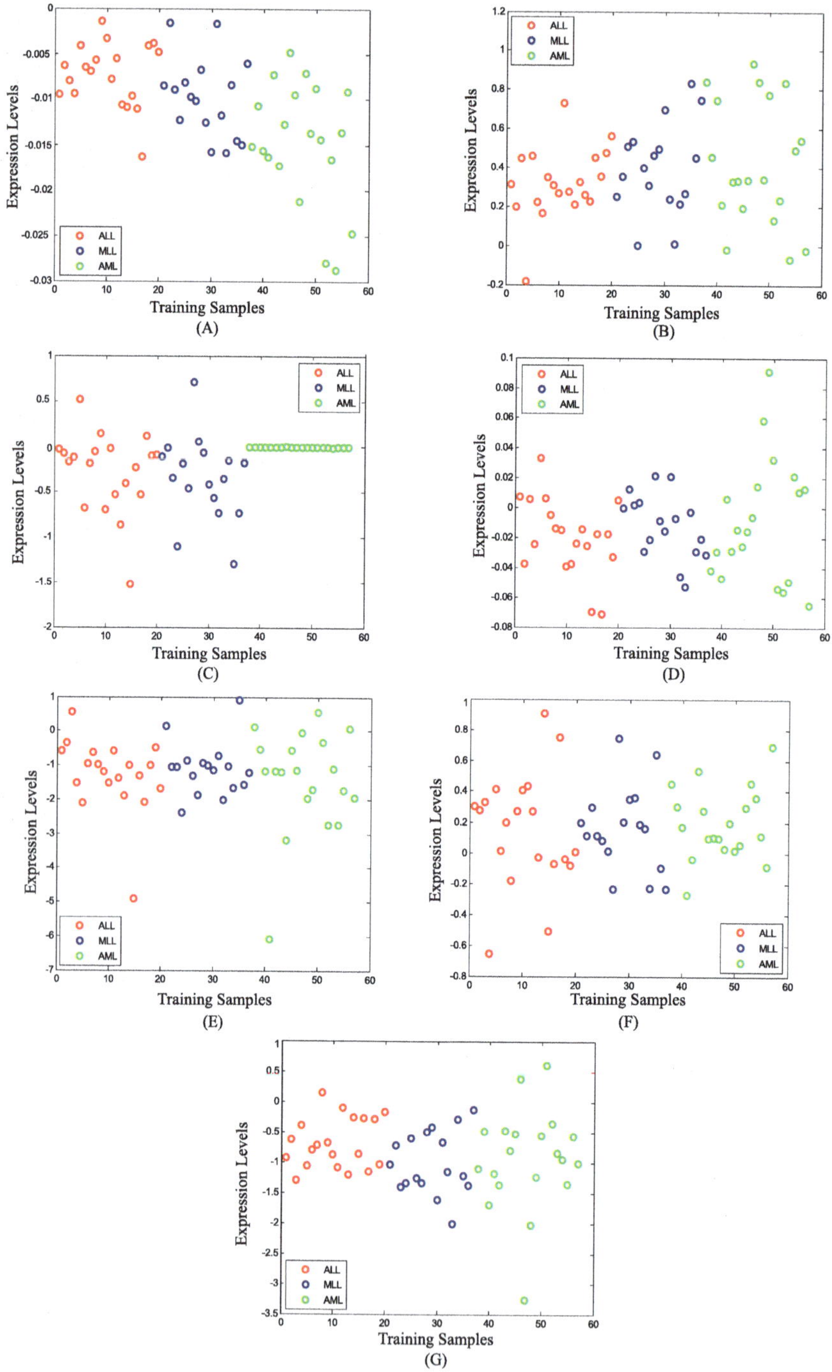

Fig. 4.25 Scatter plot of selected 7 genes of MLL data using PSO-based wrapper method. (**a**) 31351_at, (**b**) 31353_f_at, (**c**) 31412_at, (**d**) 31473_s_at, (**e**) 31553_at, (**f**) 31686_at, (**g**) 31704_at

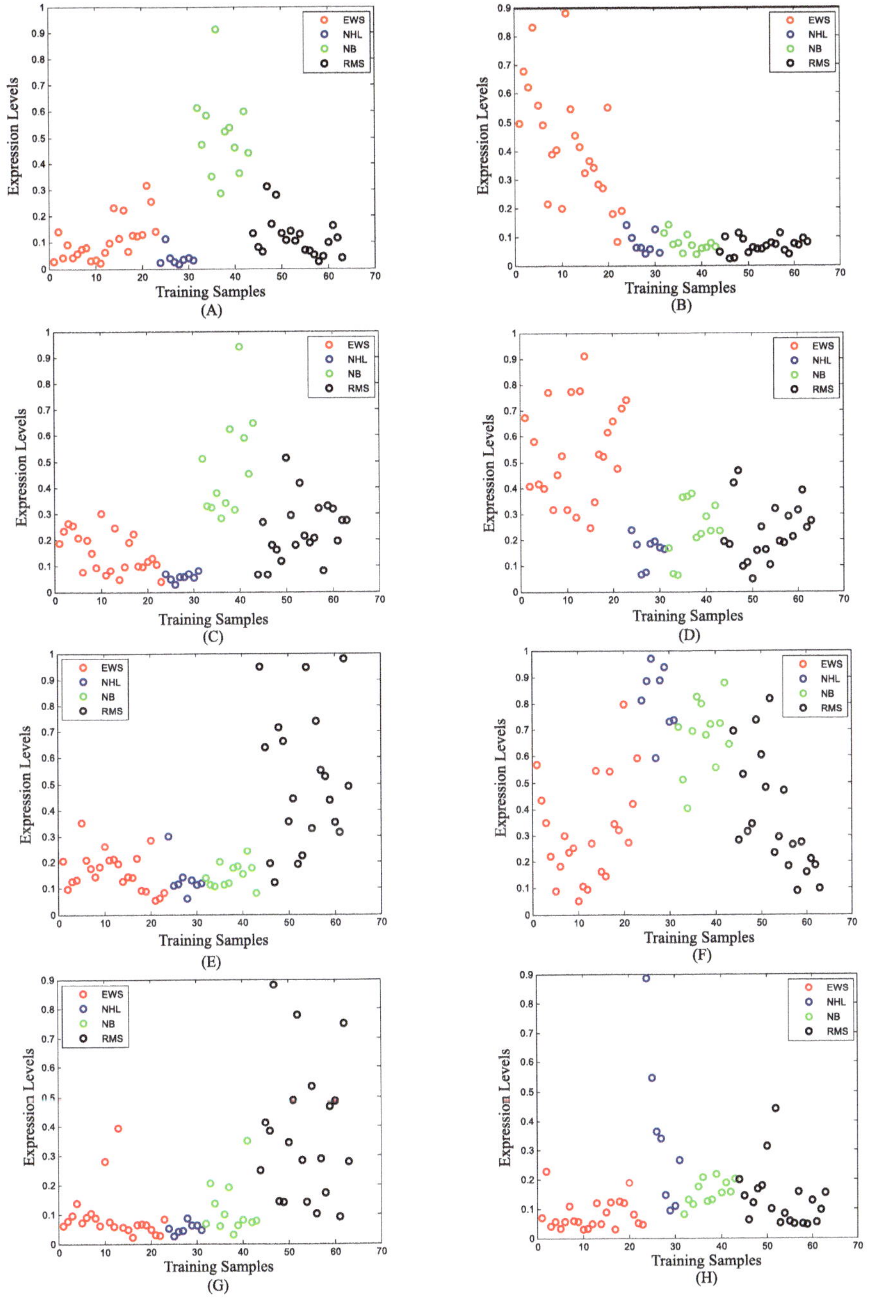

Fig. 4.26 Scatter plot of 1 to 8 genes of selected 14 genes of SRBCT data using PSO-based wrapper method. (**a**) Image ID 812105. (**b**) Image ID 814260. (**c**) Image ID 377048. (**d**) Image ID 491565. (**e**) Image ID 839552. (**f**) Image ID 789204. (**g**) Image ID 813841. (**h**) Image ID 78104

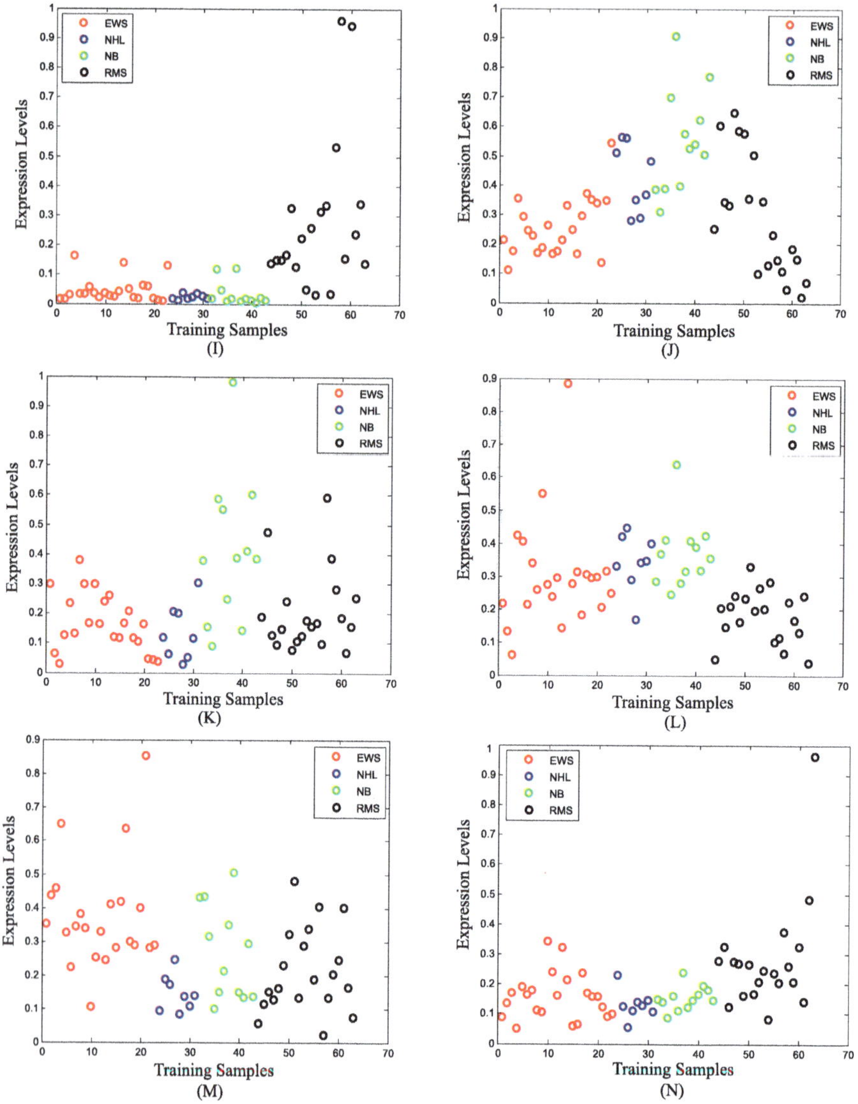

Fig. 4.27 Scatter plot of 9 to 14 genes of selected 14 genes of SRBCT data using PSO-based wrapper method. (**i**) Image ID 1409509. (**j**) Image ID 51041. (**k**) Image ID 364510. (**l**) Image ID 950367. (**m**) Image ID 75923. (**n**) Image ID 753620

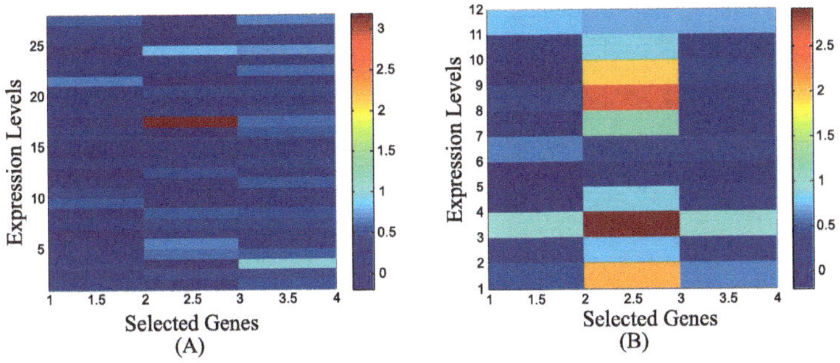

Fig. 4.28 Pseudo-color image of selected 3 genes of ALL_AML data using PSO-based wrapper method. (**a**) ALL. (**b**) AML

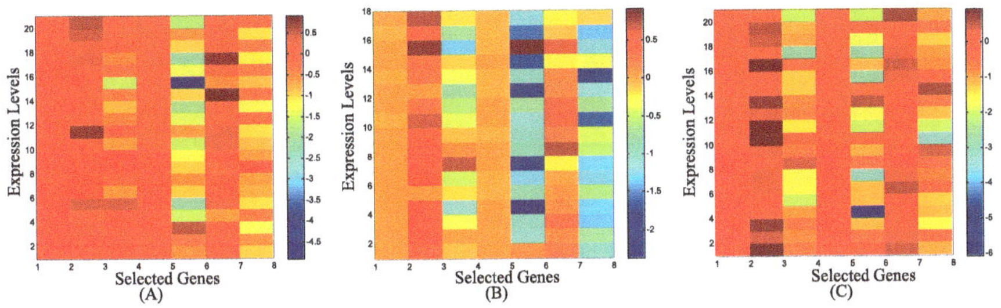

Fig. 4.29 Pseudo-color image of selected 7 genes of MLL data using PSO-based wrapper method. (**a**) ALL. (**b**) AML. (**c**) MLL

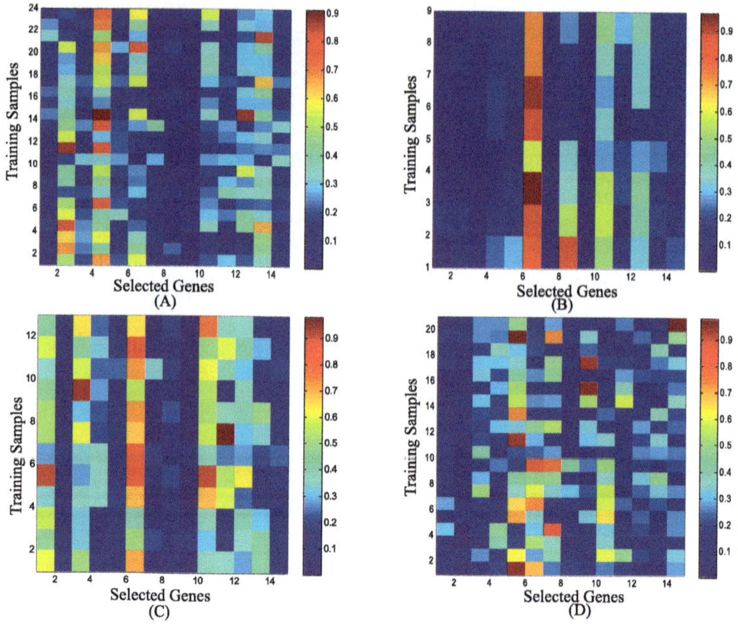

Fig. 4.30 Pseudo-color image of selected 14 genes of SRBCT data using PSO-based wrapper method. (**a**) EWS. (**b**) NHL. (**c**) NB. (**d**) RMS

References

1. Lazar C, Taminau J, Meganck S, Steenhoff D, Coletta A, Molter C, de Schaetzen V, Duque R, Bersini H, Nowe A (2012) A survey on filter techniques for feature selection in gene expression microarray analysis. IEEE/ACM Trans Comput Biol Bioinf 9:1106
2. Shen Q, Shi W-M, Kong W (2008) Hybrid particle swarm optimization and tabu search approach for selecting genes for tumor classification using gene expression data. Comput Biol Chem 32:53
3. Lu Z, Xu Q, Al-Rajab M, Chiazor L (2021) Machine learning in cancer research with applications in colon cancer and big data analysis. IGI Global, Hershey
4. Mohamad MS, Omatu S, Deris S, Yoshioka M (2011) A modified binary particle swarm optimization for selecting the small subset of informative genes from gene expression data. IEEE Trans Inf Technol Biomed 15:813
5. Chen D, Liu Z, Ma X, Hua D (2005) Selecting genes by test statistics. J Biomed Biotechnol 2005:132
6. Furey TS, Cristianini N, Duffy N, Bednarski DW, Schummer M, Haussler D (2000) Support vector machine classification and validation of cancer tissue samples using microarray expression data. Bioinformatics 16:906
7. Xiong M, Li W, Zhao J, Jin L, Boerwinkle E (2001) Feature (gene) selection in gene expression-based tumor classification. Mol Genet Metab 73:239
8. Dashtban M, Balafar M (2017) Gene selection for microarray cancer classification using a new evolutionary method employing artificial intelligence concepts. Genomics 109:91

9. Liu H, Zhao Z (2012) Manipulating data and dimension reduction methods: Feature selection. In: Computational complexity: theory, techniques, and applications. Springer, New York, pp 1790–1800
10. Liu H, Motoda H, Setiono R, Zhao Z (2010) Feature selection: An ever evolving frontier in data mining. In: Feature selection in data mining. PMLR, pp 4–13
11. Kar S, Sharma KD, Maitra M (2015) A comparative study on gene ranking and classification methods using microarray gene expression profiles. IET, pp 103–105
12. Ross SM (2020) Introduction to probability and statistics for engineers and scientists. Academic Press, Cambridge
13. Welch BL (1947) The generalization of 'STUDENT'S'problem when several different population varlances are involved. Biometrika 34:28
14. Chu F, Jin G, Wang L (2005) Cancer diagnosis and protein secondary structure prediction using support vector machines. In: Support vector machines: theory and applications, vol 177. Springer Science & Business Media, pp 343–363
15. Pearson K (1900) X. On the criterion that a given system of deviations from the probable in the case of a correlated system of variables is such that it can be reasonably supposed to have arisen from random sampling. London, Edinburgh Dublin Philos Mag J Sci 50:157
16. Liu H, Setiono R (1995) Chi2: Feature selection and discretization of numeric attributes. In: Proceedings of 7th IEEE International Conference on Tools with Artificial Intelligence. IEEE, Piscataway, pp 388–391
17. Liu H, Li J, Wong L (2002) A comparative study on feature selection and classification methods using gene expression profiles and proteomic patterns. Genome Inf 13:51
18. Dash M, Liu H (1997) Feature selection for classification. Intell Data Anal 1:131
19. Guyon I, Elisseeff A (2003) An introduction to variable and feature selection. J Mach Learn Res 3:1157
20. Chandrashekar G, Sahin F (2014) A survey on feature selection methods. Comput Electr Eng 40:16
21. De La Iglesia B (2013) Evolutionary computation for feature selection in classification problems. Wiley Interdiscip Rev Data Min Knowl Discovery 3:381
22. Li Y, Zhang S, Zeng X (2009) Research of multi-population agent genetic algorithm for feature selection. Expert Syst Appl 36:11570
23. Derrac J, García S, Herrera F (2009) A first study on the use of coevolutionary algorithms for instance and feature selection. In: Hybrid artificial intelligence systems. Proceedings of the 4th International Conference, HAIS 2009, Salamanca, June 10–12, 2009. Springer, Berlin, pp 557–564
24. Li R, Lu J, Zhang Y, Zhao T (2010) Dynamic Adaboost learning with feature selection based on parallel genetic algorithm for image annotation. Knowl Based Syst 23:195
25. Chen D, Chan KC, Wu X (2008) Gene expression analyses using genetic algorithm based hybrid approaches. In: 2008 IEEE Congress on Evolutionary Computation. IEEE World Congress on Computational Intelligence. IEEE, Piscataway, pp 963–969
26. Lin F, Liang D, Yeh C-C, Huang J-C (2014) Novel feature selection methods to financial distress prediction. Expert Syst Appl 41:2472
27. Jeong Y-S, Shin KS, Jeong MK (2015) An evolutionary algorithm with the partial sequential forward floating search mutation for large-scale feature selection problems. J Oper Res Soc 66:529
28. Eberhart R, Kennedy J (1995) A new optimizer using particle swarm theory. In: MHS'95. Proceedings of the Sixth International Symposium on Micromachine and Human Science. IEEE, Piscataway, pp 39–43
29. Li S, Wu X, Tan M (2008) Gene selection using hybrid particle swarm optimization and genetic algorithm. Soft Comput 12:1039
30. Unler A, Murat A (2010) A discrete particle swarm optimization method for feature selection in binary classification problems. Eur J Oper Res 206:528
31. Xue B, Zhang M, Browne WN (2012) Particle swarm optimization for feature selection in classification: A multi-objective approach. IEEE Trans Cybern 43:1656

32. Agrafiotis DK, Cedeno W (2002) Feature selection for structure- activity correlation using binary particle swarms. J Med Chem 45:1098
33. Shi Y, Eberhart R (1998) A modified particle swarm optimizer. In: 1998 IEEE International Conference on Evolutionary Computation Proceedings. IEEE World Congress on Computational Intelligence (Cat. No. 98TH8360). IEEE, Piscataway, pp 69–73
34. Shi Y, Eberhart RC (1998) Parameter selection in particle swarm optimization. In: Evolutionary Programming VII: 1998 Proceedings of the 7th International Conference, EP98 San Diego, CA, March 25–27. Springer, Berlin, pp 591–600
35. Kar S, Sharma KD, Maitra M (2016) A particle swarm optimization based gene identification technique for classification of cancer subgroups. In: 2016 2nd International Conference on Control, Instrumentation, Energy & Communication (CIEC). IEEE, Piscataway, pp 130–134
36. Vapnik V (1999) The nature of statistical learning theory. Springer Science & Business Media, Berlin
37. Kumar S (2004) Neural networks: a classroom approach. Tata McGraw-Hill Education, New York
38. Song Y, Huang J, Zhou D, Zha H, Giles CL (2007) Iknn: Informative k-nearest neighbor pattern classification. In: European Conference on Principles of Data Mining and Knowledge Discovery. Springer, Berlin, pp 248–264
39. PEH D, Duda RO, Hart PE, Stork DG (2001) Pattern classification. Wiley
40. Kar S, Sharma KD, Maitra M (2014) Diagnostic prediction of multi-class cancer using SVM and nearest neighbor classifier. In: Proceedings of the 2014 International Conference on Control, Instrumentation, Energy and Communication (CIEC). IEEE, Piscataway, pp 636–640

Chapter 5
Intelligent Computing Approach in Gene Evaluation for Carcinogenic Disease Detection

Abstract This chapter addresses the challenge of identifying a small subset of crucial genes from microarray data, given the limited number of effective samples compared to the vast number of genes. The work suggests a novel method combining adaptive k-nearest neighbor-based gene selection with Particle Swarm Optimization (PSO) to obtain accurate identification of cancer subtypes. The aim is to identify a small yet informative set of genes that can be properly classified. In order to effectively explore the right neighborhood and accurate classification, the suggested method incorporates a heuristics for calculating the optimal value of k. Three benchmark microarray data sets namely SRBCT, ALL_AML, and MLL are used to test the suggested gene election method. The method's effectiveness is demonstrated by the results, which include high classification accuracy on blind test samples, the number of informative genes identified, and computational efficacy. Furthermore, other classifiers, such as Support Vector Machine (SVM), are used to assess the usefulness and universal properties of the identified genes. The results confirm that the suggested methodology is resilient and useful in the field of carcinogenic disease classification.

5.1 Introduction

It has already been demonstrated that the identification of distinct cancer subgroups using microarray gene expression profiles is crucial for the diagnosis and prognosis of cancer. Numerous PSO-based wrapper approaches for gene selection have grown in prominence recently [1–4] because of their straightforward mathematical foundation. It is also computationally inexpensive in terms of speed and memory concern [5].

Researchers have used a few heuristics-based wrapping techniques on the fewer genes in the literature [1–3]. That essentially indicates a significant portion of genes have been completely ignored while a small number of genes have been assumed to have an active role in the heuristic method. However, the lacunae of the presently available approaches that have been identified are enumerated below.

K. Das Sharma et al., *Intelligent Computing in Carcinogenic Disease Detection*, Computational Intelligence Methods and Applications,
https://doi.org/10.1007/978-981-97-2424-6_5

- More numbers of genes should take part in the heuristic-based wrapper approach.
- Blind testing on the data set is also needed to evaluate the classification accuracy of the identified genes.
- Further, the usefulness of the selected genes are supposed to be reconfirmed by using a different classifier.

In order to effectively classify the cancer subgroups, it is important to find a small subset of useful genes from the collection of all genes. In order to achieve this, the whole gene space specified by the microarray data set has been searched using the stochastic optimization approach known as PSO to find the informative genes. Additionally, various classifiers, including SVM and k-NN, are used in this chapter to distinguish between different carcinoma subgroups. In order to reduce the dimensionality of the microarray data, the suggested method essentially does not use a gene ranking mechanism, i.e., filter method-based approach. Instead, each potential solution vector has been given a common threshold value. A specific collection of genes has either been approved to be retained in the relevant pool or has been excluded depending on that particular threshold value. To categorize cancer subgroups using SVM and k-NN, all the genes in the microarray have been used together, and informative genes have been chosen.

The nonparametric k-NN classifier is the foundation of this chapter. Compared to other complex classifiers, it provides greater performance [6–8]. The difficulty, though, lies in selecting the right neighborhood$k$ number for a k-NN-based multi-class classification problem. The classifier may not have the necessary robustness if the value of k is too low [9]. On the other hand, if the value of k is too large, it could result in subpar classification results [10]. As a result, a more reliable classifier may be achieved by selecting a suitable value of k that would direct the search into the appropriate neighborhood and allow for majority voting to determine the classifier's decision. In this work, a heuristic for choosing the right value of k has been proposed and is interspersed with the PSO-based gene selection process in such a way that selection and classification processes can adaptively evolve with the value of k toward improving classification accuracy with a small subset of significant genes.

Additionally, there ought to be a universal characteristic in the chosen group of genes. Actually, this means that if the chosen subset is necessary and adequate, it would generate high classification accuracy when used with any other classifier. Thus, the classification accuracy has also been assessed using the SVM classifier with various kernels in order to assess this general characteristic.

However, in a nutshell, this chapter elaborates the following work. A PSO-based wrapper strategy utilizing adaptive k-NN technique has been developed to select a subset of useful genes for accurate classification of cancer subgroups. Further, SVM classifiers with various kernels have been used to evaluate the universal characteristic of the chosen gene subset. It has been discovered that the hypothesis ensures high classification accuracy with a minimal number of useful genes.

5.2 Adaptive k-Nearest Neighborhood Technique

In this chapter, PSO has been used to conduct a comprehensive search over the whole microarray gene set in order to identify the pertinent genes that are strong candidates for classification across various cancer subgroups. In order to increase classification accuracy, the PSO-based gene selection technique has been included into the selection methodology for the value of$k$ with reference to the k-NN classifier. Here, the proposed methodology has additionally been strengthened using the K-fold cross-validation technique. In order to achieve improved classification accuracy while using fewer relevant genes, the suggested technique has adaptively evolved both the gene selection and k selection methodologies [11]. Moreover, the universal characteristic of the identified genes has been reconfirmed by utilizing different kernels of SVM classifier. The block diagram of proposed PSO-adaptive k-NN-based gene selection methodology has been shown in Fig. 5.1.

5.2.1 PSO-Based Adaptive k-Nearest Neighborhood Technique for Gene Evaluation

The total data sample is separated into K-folds in this method. The remaining (K-1) folds are used as training samples, and one fold of those K folds is used as validation data for testing. This same method is done K times, using exactly one of each fold's test data. This particular method considers the dimension of each particle to be equal to the number of genes for each sample in the microarray gene expression data. Here, each particle has been initialized with random numbers between 0 and 1. Then a threshold ϵ is utilized to construct the gene selection flag using Eq. (4.21). Each subsequent particle is represented as a binary gene selection flag, with "1" denoting that the particular gene is present in the selected feature subset and "0" denoting that it is absent from the selected subset. Each particle with the subset of selected genes is then utilized to calculate the cross-validation (CV) training accuracy (CV training accuracy), cross-validation test accuracy (CV test accuracy), and fitness function.

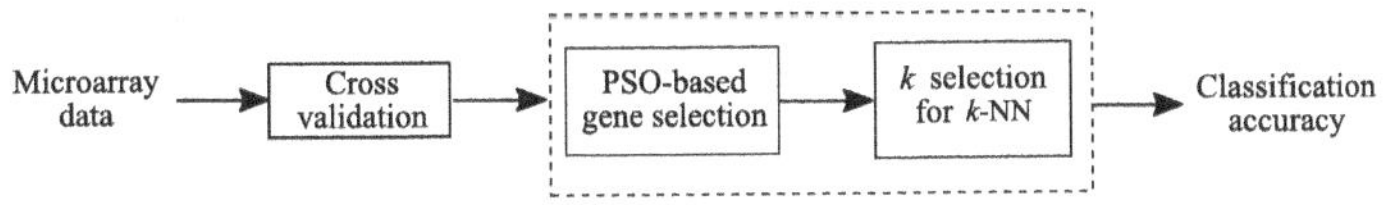

Fig. 5.1 Block diagram of the proposed methodology

For each training and testing phase, k for k-NN classifier is varied from $k_{initial}$ to k_{final}. The CV training accuracy and CV test accuracy for the present case are calculated as:

$$CV\ training\ accuracy = \frac{1}{K}\sum_{m=1}^{K} Tr_m \tag{5.1}$$

$$CV\ test\ accuracy = \frac{1}{K}\sum_{m=1}^{K} Ts_m \tag{5.2}$$

where Tr and Ts are training and test accuracies for each fold, respectively. For each k, k number of single estimations (SE) are calculated as:

$$SE\,(k) = \frac{1}{2}\sum_{k=k_{initial}}^{k_{final}} \left(\frac{1}{K}\sum_{m=1}^{K} Tr_m + \frac{1}{K}\sum_{m=1}^{K} Ts_m \right) \tag{5.3}$$

From the k number of single estimations, an optimal value is selected as: $max\,(SE\,(k))$. This optimal value is named as Cross-Validation accuracy (CV accuracy) and corresponding k value for k-NN is selected as k_{opt}. The CV accuracy and k_{opt} are evaluated for each particle in each iteration of the proposed algorithm. The simplified diagram of the proposed methodology is shown in Fig. 5.2.

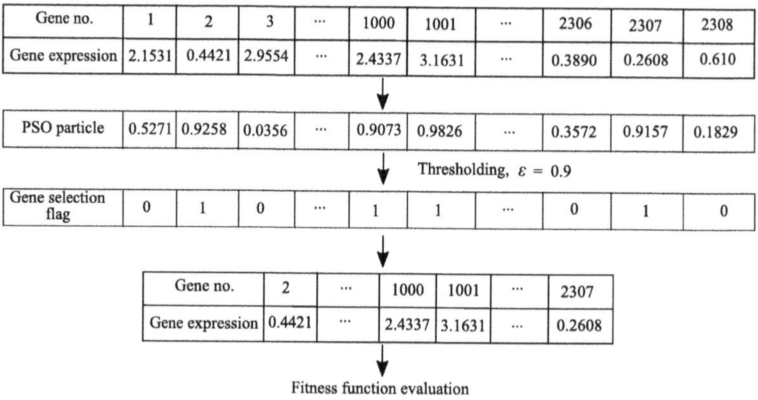

Fig. 5.2 Simplified diagram of the proposed methodology

5.2.2 Fitness Function

The goals of the suggested methodology are dual. The first goal is to improve the cancer subgroup classification accuracy, and the second goal is to find fewer informative genes that also contribute to improved classification accuracy. As a result, the fitness function in the suggested technique is built as:

$$FF = \omega_1 \times A_c + \omega_2 \times \frac{1}{|N_G|} \tag{5.4}$$

where A_c= CV accuracy and $|N_G|$ is the number of genes in the candidate gene subset. ω_1 and ω_2 represent two different weights associated with two different objectives. In the proposed technique, the two weights are selected as $\omega_1 = \omega_2 = 0.5$ for assigning equal importance to different objectives. The proposed PSO-adaptive *k*-NN-based gene selection technique is shown in Algorithm 5.1.

5.3 Experimental Study

On the provided data sets, the proposed technique is used to conduct numerous thorough analyses individually. After all the analysis was completed, an average result was produced. The proposed method is a stochastic strategy, thus the determination of the average result is crucial. A few key factors, including the amount of informative genes chosen, CV accuracy, training accuracy, and test accuracy on blind test samples, have been taken into account to gauge the effectiveness of the suggested technique. A superior experimental outcome requires high blind testing accuracy with fewer informative genes [11].

On each of the three benchmark data sets, the analysis has been run ten times in this study. The results of this ten independent analysis are shown in Tables 5.1, 5.2, and 5.3 for SRBCT, ALL_AML, and MLL data sets, respectively. Furthermore, highlighted shells emphasize the best outcomes. This demonstrates that the suggested methodology selects an ideal subset of informative genes effectively.

Relation of CV accuracy, number of genes, fitness function, and the value of *k* with the progress of iteration are shown in Figs. 5.3, 5.4, and 5.5, respectively. While gene numbers tend to move downward with iteration, CV accuracy trends upward. Since gene numbers and CV accuracy are combined to create the fitness function, the fitness function's trend is likewise rising as iteration progresses. This tendency implies that the suggested strategy finds a limited group of informative genes that can improve classification performance. However, each problem is different in terms of how many iterations are required to arrive at an ideal solution [12]. Therefore, based on the dimensionality of each data set, a varied number of iterations is used.

The CV training accuracy and CV test accuracy corresponding to the optimal solution are also shown in Figs. 5.6, 5.7, and 5.8 for the SRBCT, ALL_AML, and

Algorithm 5.1 Proposed PSO-adaptive k-NN-based gene selection Algorithm

```
for each particle do
    initialize position
    initialize velocity
end for
repeat
//gene selection flag//
for each particle do
    for each dimension do
        construct training samples using Eq. (4.21)
    end for
    repeat
    //calculation of CV accuracy and k_opt//
    for each k ∈ k_initial, ..., k_final do
        train the classifier using (K − 1) folds
        test the classifier using K^th fold
    end for
    until each of the K fold used exactly ones as test fold
    for each k ∈ k_initial, ..., k_final do
        evaluate CV training accuracy using Eq. (5.1)
        evaluate CV test accuracy using Eq. (5.2)
        evaluate SE (k) using Eq. (5.3)
    end for
    evaluate CV accuracy as: max {SE (k)}
    select the corresponding value of k as k_opt
    select |N_G| corresponding to CV accuracy
    //fitness function//
    evaluate fitness function using Eq. (5.4)
    //personal best and global best particle//
    evaluate personal best (p_best) position using Eq. (4.14)
end for
evaluate global best (g_best) position using Eq. (4.15)
//update position and velocity//
calculate new velocity and new position using Eqs. (4.16) and (4.18)
until convergence criteria is satisfied
```

Table 5.1 Experimental result for each run on SRBCT data set

Run	No. of Genes	Index no. of selected genes	Cross-validation Acc.	*k* for *k*-NN	Training Acc.	Test Acc.	Computing Time (h)
1	13	187,359,390,742,1051,1074,1374, 1738,1841,1842,1947,1956,1980	98.0159	6	98.4127	90	5.5859
2	13	107,123,246,518,714,742,1181, 1206,1286,1343,1954,2097,2106	99.2063	4	100	90	2.8632
3	5	246,257,742,1076,2103	93.2540	4	93.6508	90	2.8098
4	10	187,231,246,356,428,534,650,742, 758,840	95.6349	6	95.2381	90	2.7993
5	6	742,1003,1386,2046,2099,2157	98.0159	4	100	100	2.7956
6	9	125,187,246,362,742,783,1496, 1764,2040	96.8254	3	96.8254	90	2.8111
7	7	742,1601,1645,1932,1955, 2046,2144	93.2540	3	100	100	2.8202
8	7	84,251,742,1032,1074,1645,1783	94.4444	4	100	100	2.8098
9	9	255,694,806,1003,1117,1319,1477, 2046,2113	96.8254	3	100	95	2.8017
10	6	246,841,1263,1531,1955,1645	94.8413	3	96.8254	95	2.7981
Avg.	8.5	–	96.0318	–	98.0952	94	3.0895
Std. Dev.	2.8382	–	2.0746	–	2.3424	4.5947	0.8774

Table 5.2 Experimental result for each run on ALL_AML data set

Run	No. of Genes	Index no. of selected genes	Cross-validation Acc.	k for k-NN	Training Acc.	Test Acc.	Computing Time (h)
1	3	804,1249,1334	97.4206	4	100	94.1176	2.8065
2	3	804,1882,4052	95.8868	4	100	97.0588	2.7906
3	2	1882,4108	95.8333	3	100	94.1176	2.4086
4	4	804,1882,4052,2642	100	4	100	97.0588	2.8065
5	4	187,1096,1882,3747	95.1389	3	97.3684	94.1176	2.8065
6	2	2642,4052	97.9701	6	97.3684	76.4706	2.7973
7	3	804,1882,2642	98.6111	4	100	97.0588	2.7956
8	2	1882,4052	94.5513	6	94.7368	94.1176	2.6201
9	2	1882,2642	97.2222	4	100	94.1176	2.4086
10	2	804,1882	98.6111	4	97.3684	97.0588	2.7906
Avg.	2.7	–	97.1245	–	98.6842	93.5294	2.7031
Std. Dev.	0.8233	–	1.7405	–	1.8608	6.1695	0.1651

Table 5.3 Experimental result for each run on MLL data set

Run	No. of Genes	Index no. of selected genes	Cross-validation Acc.	k for k-NN	Training Acc.	Test Acc.	Computing Time (h)
1	4	2835,4341,8165,8937	98.6842	6	100	93.3333	7.1088
2	3	4341,2028,12418	86.4035	6	94.7368	93.3333	7.2089
3	4	3479,657,841,7930	76.3158	3	91.2281	86.6667	7.2917
4	4	841,7930,12418,4341	88.1579	4	96.4912	93.3333	7.3956
5	4	841,7930,8937,12418	92.5439	4	100	100	7.1488
6	4	4341,7930,2028,841	87.7193	6	91.2281	86.6667	7.1667
7	3	657,7136,12418	85.0877	4	96.4912	86.6667	7.2761
8	2	841,7136	83.33	3	91.2281	86.6667	7.1962
9	3	4937,4982,7136	89.0351	4	96.4912	93.3333	7.2058
10	2	7136,4341	92.1053	4	92.9825	86.6667	7.1477
Avg.	3.3	–	87.9383	–	95.0877	90.6667	7.2146
Std. Dev.	0.8233	–	5.9828	–	3.3898	4.6613	0.0851

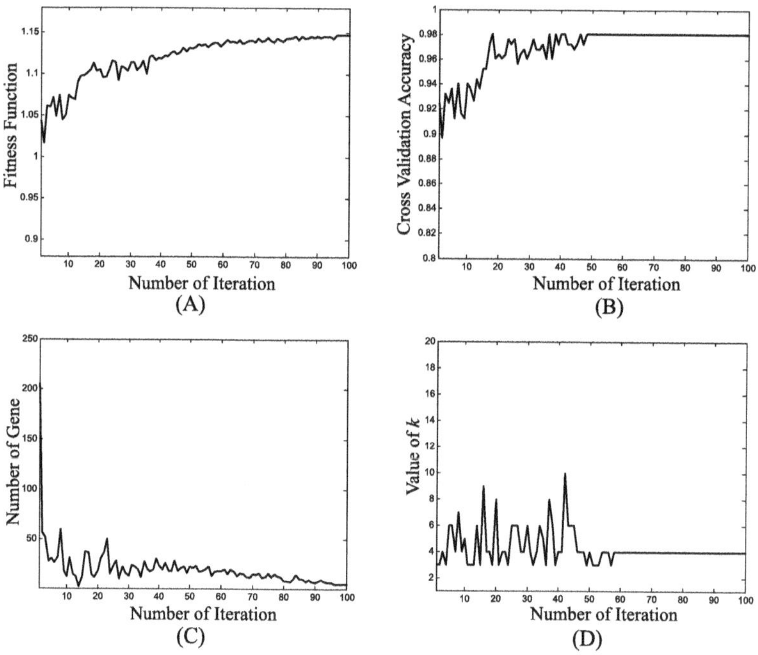

Fig. 5.3 Result after 100 PSO iterations on SRBCT data (**a**) fitness function is increasing with the number of iterations, (**b**) cross-validation accuracy initially increases with number of iterations and then gets saturated, (**c**) number of genes is decreasing with number of iterations, and (**d**) average value

MLL data sets, respectively. It is observed from the graphical representation that value of CV accuracy is maximum for $k = 4$, and therefore $k_{opt} = 4$ is selected.

The scatter plot of SRBCT data is shown in Fig. 5.9 where each panel corresponds to one selected gene. The red, blue, green, and black colors, respectively, correspond to EWS, NHL, NB, and RMS type of SRBCTs. It is observed that the gene AF1Q is highly expressed for NB, gene SGCA is highly expressed for RMS, gene EHD1 is highly expressed for NHL and moderately expressed for a few cases of RMS, gene EST is highly expressed for RMS, gene image ID 234376 is highly expressed for EWS and gene image ID 244637 is highly expressed for NB.

The scatter plot of ALL_AML data is shown in Fig. 5.10 where each panel corresponds to one selected gene. The red and blue colors correspond to ALL and AML types, respectively. It is noticed that gene HG1612-HT1612_at is highly expressed for ALL, gene M27891_at is highly expressed for AML, and gene X04085_rna1_at is highly expressed for AML and for some cases moderately expressed for ALL.

The scatter plot of MLL data is shown in Fig. 5.11 where each panel corresponds to one selected gene. The red, blue, and green colors correspond to ALL, MLL, and AML types, respectively. It is also observed that gene 33054_at is moderate

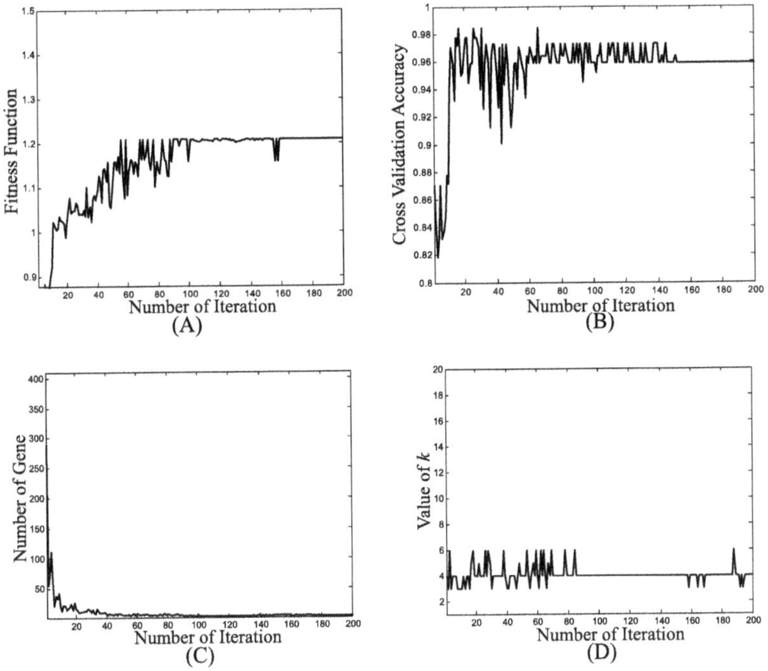

Fig. 5.4 Result after 200 PSO iterations on ALL_AML data. (**a**) Fitness function initially increases with the number of iterations and then gets saturated (**b**) Number of genes is decreasing with number of iterations, (**c**) Cross-validation accuracy initially increases with number of iterations and then gets saturated, and (**d**) average value of k remains almost constant and finally gets saturated at a value of 4

to highly expressed for AML, gene 34306_at is highly expressed for MLL, gene 37710_at is moderate to highly expressed for ALL and for a few cases of MLL, and gene 266_s_at is highly expressed for ALL and moderately expressed for a few cases of MLL.

The pseudo color images for the three data sets are shown in Figs. 5.12, 5.13, and 5.14, respectively, where each panel corresponds to one class. Blue to red colors represent low to high expression levels of the training samples. In Fig. 5.12, the EWS panel reveals that moderate to high upregulation of Image ID 234376 and moderate upregulation of Image ID 244637 and downregulation for other four genes can signal EWS. The NHL panel represents high upregulation of EHD1 and downregulation for other five genes can indicate the presence of NHL. The NB panel suggests that high upregulation of AF1Q and Image ID 244637 and downregulation for other four genes are indicators of NB, and RMS panel suggests that moderate to high upregulation of SGCA and EST and moderate upregulation for a few cases of EHD1 and Image ID 244637 and downregulation of other two genes can indicate the presence of RMS.

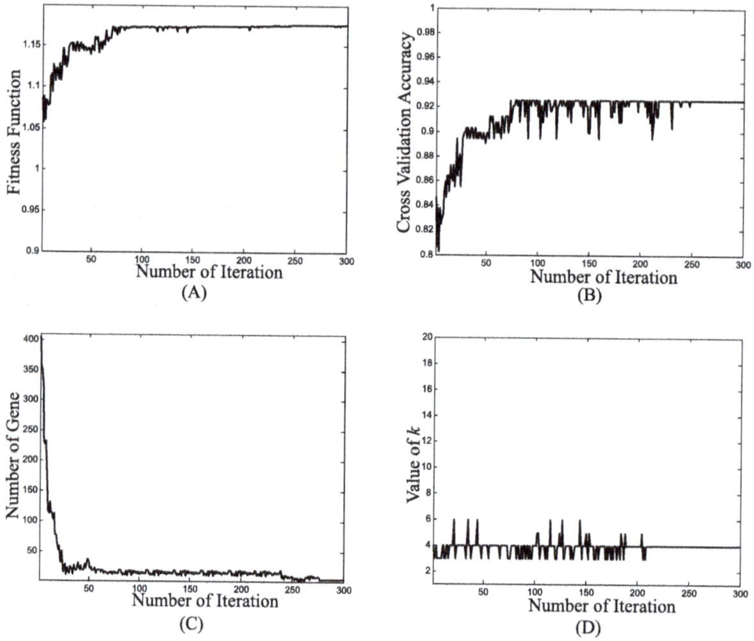

Fig. 5.5 Result after 300 PSO iterations on MLL data. (**a**) Fitness function initially increases with the number of iterations and then gets saturated, (**b**) Number of genes is decreasing with number of iterations, (**c**) Cross-validation accuracy initially increases with number of iterations and then gets saturated, and (**d**) average value of *k* remains almost constant and finally gets saturated at a value of 4

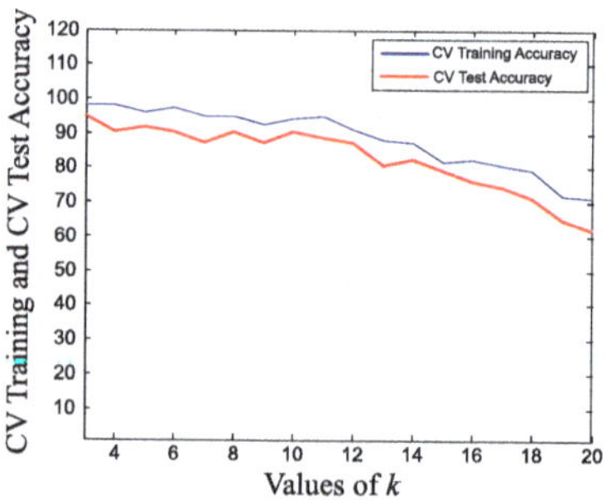

Fig. 5.6 CV Training and CV Test Accuracy vs. Values of *k* for SRBCT data set

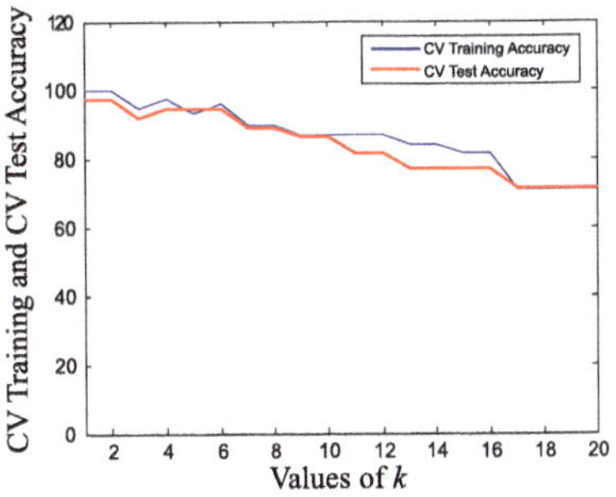

Fig. 5.7 CV Training and CV Test Accuracy vs. Values of *k* for ALL_AML data set

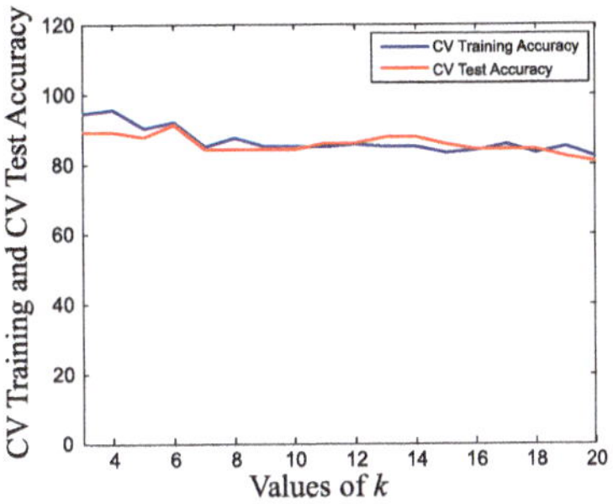

Fig. 5.8 CV Training and CV Test Accuracy vs. Values of *k* for MLL data set

In Fig. 5.13, the ALL panel reveals that moderate to high upregulation of HG1612-HT1612_at and moderate upregulation of X04085_rna1_at and downregulation for other genes can signal ALL. The AML panel suggests that high upregulation of X04085_rna1_at and moderate to high upregulation for a few cases of M27891_at and downregulation for other genes can indicate the presence of AML.

In Fig. 5.14, the ALL panel reveals that high upregulation of 266_s_at and moderate to high upregulation of 37710_at and downregulation for other two genes can signal ALL. The MLL panel suggests that high upregulation of 34306_at and moderate to high upregulation for a few cases of 37710_at and downregulation for other two genes can indicate the presence of MLL and the AML panel suggests that moderate to high upregulation for a few cases of 33054_at and downregulation for other three genes are indicator of AML.

5.4 Analysis of Experimental Results

For the diagnostic prediction of malignancies, it is essential to choose a small subset of useful genes among a vast number of genes. The diagnostic prediction of SRBCT, ALL_AML, and MLL data has thus been the subject of numerous research proposals. In contrast to past works, the current technique only identifies 6 genes for SRBCT data, 3 genes for ALL_AML data, and 4 genes for MLL data as being

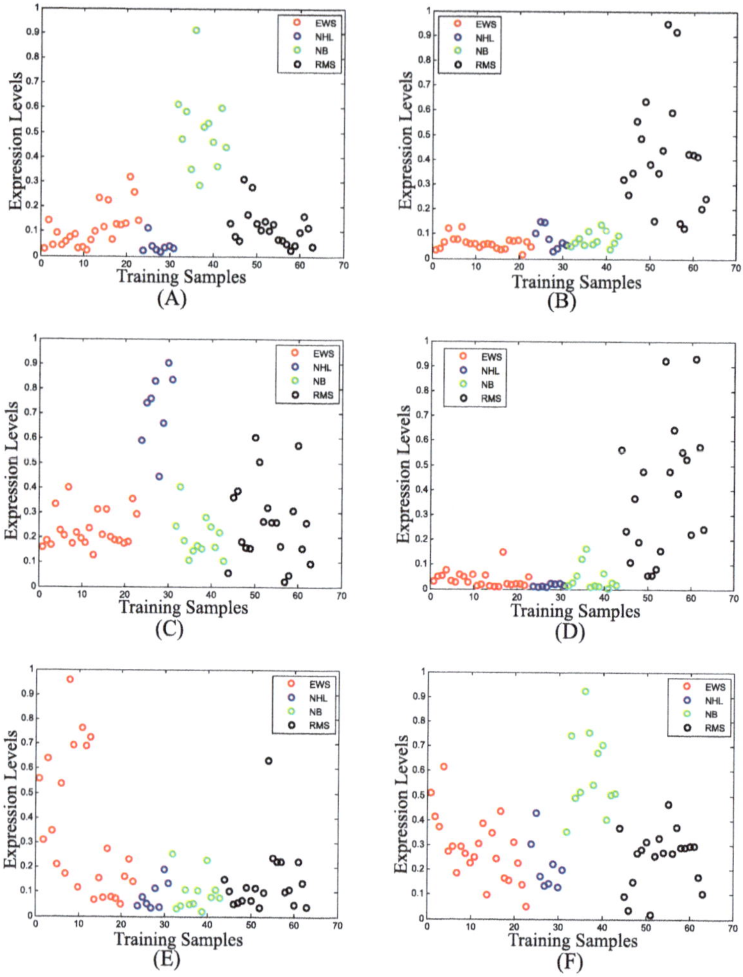

Fig. 5.9 Scatter plot of six identified genes in the training data of SRBCT data set. (**a**) Gene AF1Q (**b**) Gene SGCA (**c**) Gene EHD1 (**d**) Gene EST (**e**) Gene image ID 234376 and (**f**) Gene image ID 244637

crucial. 100% classification accuracy on blind test samples is obtained on SRBCT and MLL data whereas 97.0588% classification accuracy is obtained for ALL_AML data, i.e., only one misclassification occurred out of 34 blind test samples. The analysis of the obtained result has been discussed in the following section.

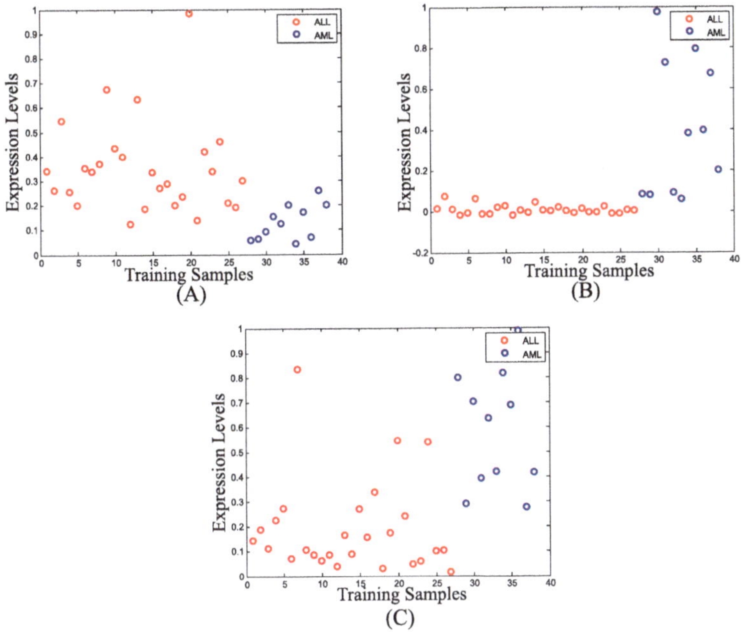

Fig. 5.10 Scatter plot of three identified genes in the training data of ALL_AML data set. (**a**) Gene HG1612-HT1612_at (**b**) Gene M27891_at, and (**c**) Gene X04085_rna1_at

5.4.1 Analysis of SRBCT Data Set

J. Khan et al. [13] used an ANN-based approach on the SRBCT data set to identify 96 relevant genes that provided 100% accuracy on test data samples. Forty three essential genes were chosen by R. Tibshirani et al. [14] using a closest shrunken centroid-based technique, resulting in 100% training and test accuracy. Feature selection multi-layered perceptron and non-Euclidean relational fuzzy c-means clustering based method were adopted by N. R. Pal et al. [16] and only 7 genes were identified as important for 100% classification accuracy on training and test samples of the same data set. Only 14 genes were identified as informative to categorize test samples with 100% accuracy using an adaptive GA and k-NN-based technique by C. P. Lee et al. [15] on the same data. A comparison of the different approaches on the SRBCT data set is shown in Table 5.4.

The best subset of selected informative genes for SRBCT data by the proposed method is shown in Table 5.5. It has been observed from Figs. 5.9 and 5.12 that AF1Q is highly expressed for NB. This particular gene was also found by J. Khan et al. [13], N. R. Pal et al. [16], L. M. Fu et al. [17], and R. Tibshirani et al. [14]. Gene Expression Omnibus (GEO) demonstrates that AF1Q discriminates between NB and Ewing family tumor (GPL96, 211071_s_at (ID_REF), GDS1713). It has been also observed that SGCA is highly expressed for RMS. The gene was

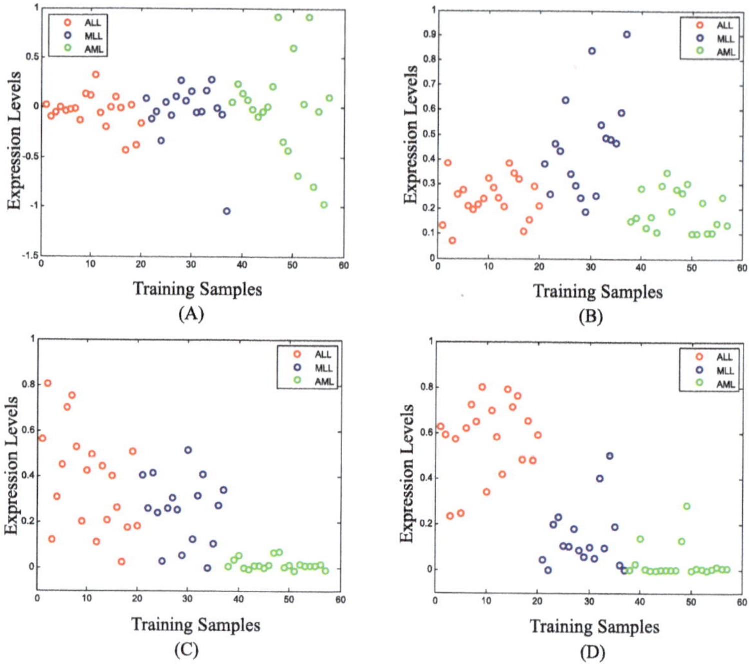

Fig. 5.11 Scatter plot of four identified genes in the training data of MLL data set. Each panel corresponds to one gene. The red, blue, and green colors correspond to ALL, MLL, and AML types, respectively. (**a**) Gene 33054_at (**b**) Gene 34306_at (**c**) Gene 37710_at, and (**d**) Gene 266_s_at

also found by J. Khan et al. [13] and N. R. Pal et al. [16]. A search in GEO profiles also showed that SGCA can discriminate RMS from Ewing's sarcoma (GPL91, 38609_at (ID_REF),GDS971). EHD1 is highly expressed for NHL and also moderately expressed for few cases of RMS. The gene was also found by N. R. Pal et al. [16]. GEO profiles also showed that EHD1 is moderately expressed for RMS (GPL91,40098_at (ID_REF),GDS971). It has been also observed that EST is highly expressed for RMS. This gene was found by J. Khan et al. [13] and GEO profiles showed that EST is highly expressed for RMS (GPL91,33961_at (ID_REF),GDS971). Furthermore, it has been observed that Image ID 234376 is highly expressed for EWS. The particular gene was also found by Bhattacharyya et al. [18]. They also observed that this gene can discriminate between EWS and other SRBCT subgroups. It has been also observed that Image ID 244637 is highly expressed for NB. Yeo et al. [19] also found image ID 244637.

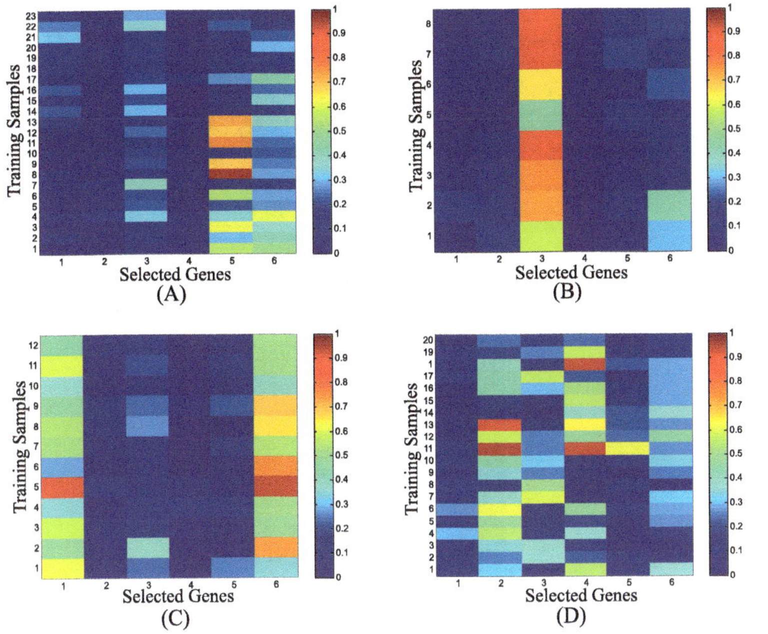

Fig. 5.12 Pseudo color image of the training data for the 4 SRBCT classes. (**a**) EWS (**b**) NHL (**c**) NB and (**d**) RMS

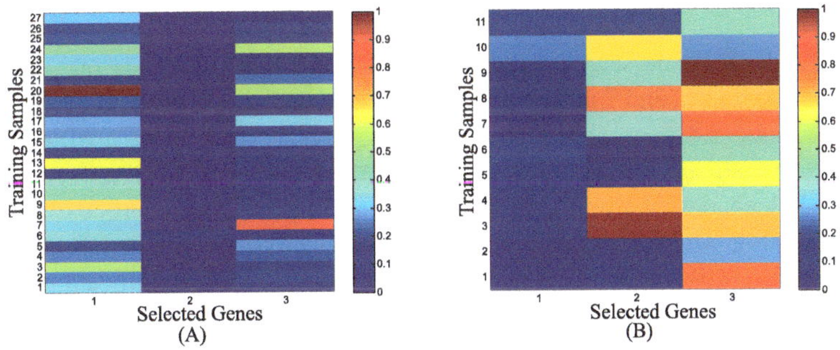

Fig. 5.13 Pseudo color image of the training data for the 2 ALL_AML classes. (**a**) ALL (**b**) AML

5.4.2 Analysis of ALL_AML Data Set

L. M. Fu et al. [17] employed recursive feature elimination and an SVM-based technique to the SRBCT and ALL AML data set. For the SRBCT data set, 19 useful genes were discovered with a test accuracy of 100%, and 4 significant genes were chosen for the ALL_AML data set, which could categorize test samples with a 97.06% accuracy. Four techniques such as stepwise, pure tabu search, pure PSO,

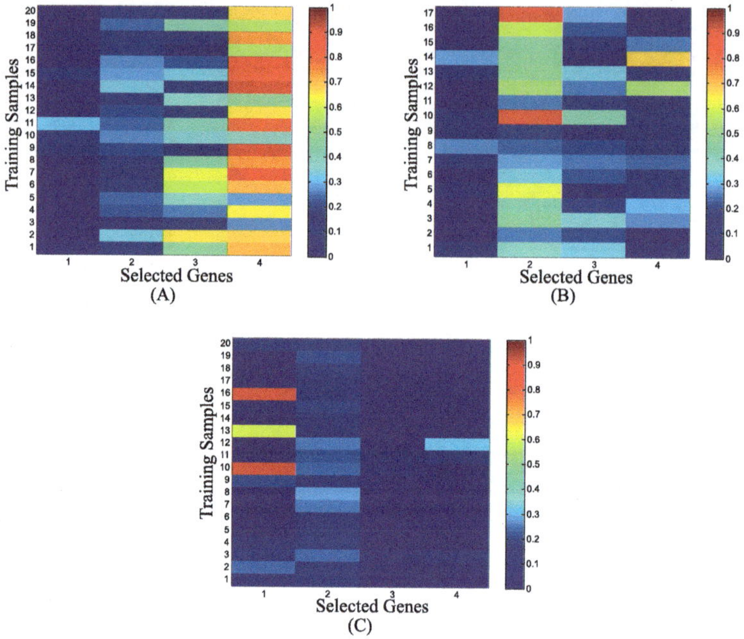

Fig. 5.14 Pseudo color image of the training data for the 3 MLL classes. (**a**) ALL (**b**) MLL and (**c**) AML

and hybrid PSO-tabu search were proposed by Q. Shen et al. [1] on ALL_AML data set and 3, 5, 7, and 7 genes were identified, respectively, with 88.14%, 94.24%, 94.19%, and 95.81% accuracy on test samples, respectively. Two methods namely partial least square variant importance in projection (PLSVIP) and partial least square independent-variable explanation gain (PLSIEG) were proposed by G. Ji et al. [21] for SRBCT and ALL AML data. Twenty-four significant genes were identified with 100% test accuracy for SRBCT data set and 9 significant genes were identified with 100% test classification accuracy for ALL AML data set by using PLSVIP technique. Similar to this, using the PLSIEG approach, 100% test accuracy was achieved utilizing only 15 and 8 chosen genes for the SRBCT and ALL AML data, respectively. P. Ganesh Kumar et al. [26] developed a novel Genetic Swarm Algorithm for ALL AML data for the accumulation of knowledge in the form of if-then rules and membership functions. Only 10 genes had been discovered to attain 100% LOOCV accuracy. Z. Zainuddin et al. [23] used a modified wavelet neural network-based technique to analyze the SRBCT and ALL_AML data sets. In order to achieve 100% 10-fold cross-validation accuracy on SRBCT data and 98.6% accuracy on ALL AML data set, just 10 genes were discovered to be crucial. A comparison of the different approaches on the ALL_AML data set is shown in Table 5.6.

Table 5.4 Comparison of the Methods on the SRBCT data set

Experiment	Method	Genes				Cross-Validation Accuracy (%)				Training Accuracy (%)	Test Accuracy (%)
J. Khan et al. [13]	ANN	96				–				100	100
R. Tibshirani et al. [14]	NSC	43				–				100	100
L. M. Fu et al. [17]	SVM-RFE	19				–				100	100
K. Yang et al. [20]		5CV		LOOCV		5 CV		LOOCV		–	–
		k-NN	SVM	*k*-NN	SVM	*k*-NN	SVM	*k*-NN	SVM		
	GS1	88	93	57	34	98	97.9	98.8	98.8	–	–
	GS2	90	99	77	96	98.1	99	98.8	100	–	–
	Chos	98	98	82	80	90.2	94.3	92.8	98.8	–	–
	F-test	90	95	89	78	98	99.2	98.8	100	–	–
N. R. Pal et al. [16]	FSMLP	7				–				100	100
	+										
	NERFCM										
G. Ji et al. [21]	PLSVIP	24				–				100	100
	PLSVEG	15				–				100	100
M. S. Mohamad et al. [3]	IBPSO	6				100				–	–
A. Sharma et al. [22]	SFS+LDA with NCC	4				–					
										100	100
	SFS+Bayes classifier	4				–				100	90
	SFS+NNC	4				–				100	95
Z. Zainuddin et al. [23]	MSFCM+WNN	10				10 CV				–	–
						100					
X. Li et al. [24]	KLLE+LLE+PCA	20				–				100	100
C. P. Lee et al. [15]	AGA+KNN	14				–				100	100
K. H. Chen et al. [25]	PSODT	–				5 CV				–	–
						92.94					
This work	Proposed technique	6				98.0159				100	100

Table 5.5 List of best subset of genes for SRBCT data

Index no. of selected genes	Image ID	Gene ID	Gene Name
742	812105	AF1Q	Transmembrane Protein
1003	796258	SGCA	"Sarcoglycan, alpha (50KD dystrophin-associated glycoprotein)"
1386	745019	EHD1	EH domain containing 1
2046	244618	EST	ESTs
2099	234376	*	Homo Sapiens mRNA: cDNA DKFZp564F112 (from clone DKFZp564F112)
2157	244637	*	Homo Sapience mRNA full length insert cDNA clone EUROIMAGE

Table 5.7 shows the best subset of selected genes for ALL_AML data set. It is found that HG1612-HT1612_at is highly expressed for ALL. This particular gene was also found by Q. Shen et al. [1] and GEO profiles showed that the gene is highly expressed for ALL (GPL1261,1435627_x_at (ID_REF),GDS4303). It is also observed that M27891_at is highly expressed for AML and this gene was also found by L. M. Fu et al. [17] and G. Ji et al. [21]. GEO profile also shows that it is highly expressed for AML (GPL8300,39689_at (ID_REF),GDS1059). Further, X04085_rna1_at is found by the present method. It has been again observed that X04085_rna1_at is highly expressed for AML and in some cases moderately expressed for ALL. Interestingly, a search in GEO profiles also showed a high expression of X04085_rna1_at for AML (GPL8300,37009_at (ID_REF),GDS1059).

5.4.3 Analysis of MLL Data Set

A. Sharma et al. [22] used successive feature selection (SFS) with linear discriminant analysis (LDA) and closest centroid classifier (NCC), successive feature selection (SFS) with Bayes classifier, and successive feature selection (SFS) with nearest neighbor classifier (NNC) approaches on both the SRBCT and MLL data sets. Only 4 discriminating genes were identified by all these three methods for SRBCT data, but on test samples, 100% classification accuracy achieved by combining SFS and LDA with NCC technique, 90% classification accuracy achieved by combining SFS and Bayes classifier technique and 95% classification accuracy achieved by combining SFS and NNC technique. By combining SFS and NNC classifier, 93% test accuracy was attained, and 4 distinguishing genes were also chosen for MLL data using the techniques suggested by A. Sharma et al. [22]. However, by utilizing other techniques suggested by A. Sharma et al. [22], 100% test accuracy for MLL data was attained. B. Chandra et al. [28] suggested a reliable Naïve-Bayes classifier (NBC)-based technique, and they reported 94.29% 10-fold

Table 5.6 Comparison of the Methods on the ALL_AML data set

Experiment	Method	Genes				Cross-Validation Accuracy (%)						Training Accuracy (%)	Test Accuracy (%)
L. M. Fu et al. [17]	SVM-RFE	4				–						100	97.06
K. Yang et al. [20]		5CV		LOOCV		5CV			LOOCV			–	–
		k-NN	SVM	*k*-NN	SVM	*k*-NN	SVM		*k*-NN		SVM		
	GS1	100	93	60	4	97.9	97.9		98.6		98.6	–	–
	GS2	85	98	10	25	97.1	97.4		98.6		98.6	–	–
	Chos	100	98	9	80	96.8	97		97.2		98.6	–	–
	F-test	96	99	25	33	97.4	97.5		98.6		98.6	–	–
Q. Shen et al. [1]	Stepwise	3				–						90.83	88.14
	Pure TS	5				–						95.83	94.24
	Pure PSO	7				–						94.75	94.19
	HPSOTS	7				–						98.08	95.81
G. Ji et al. [21]	PLSVIP	9				–						100	100
	PLSIEG	8				–						100	100
M. S. Mohamad et al. [3]	IBPSO	2				100						–	–
Z. Zainuddin et al. [23]	MSFCM+WNN	10				10 CV						–	–
						98.61							
T. T. Wong et al. [27]	Probabilistic Mechanism	–				SVM			*k*-NN			–	–
						97.38			98.21				
B. Chandra et al. [28]	RNBC	–				10 CV						–	–
						RNBC		NBC		*k*-NN			
						94.29				85.71			
								84.29					
P. Ganesh Kumar et al. [26]	GSA	10				100						–	–
This work	Proposed technique	3				95.8868						100	97.0588

Table 5.7 List of best subset of genes for ALL_AML data

Index no. of selected genes	Gene Accession Number	Gene Name
804	HG1612-HT1612_at	Macmarcks
1882	M27891_at	CST3 cystatin C (amyloid angiopathy and cerebral hemorrhage)
4052	X04085_rna1_at	Catalase (EC 1.11.1.6) 5'flank and exon 1 mapping to chromosome11, band p13 (and joined cds)

cross-validation accuracy on the ALL_AML data set and 87.14% 10-fold cross-validation accuracy on the MLL data set. K. H. Chen et al. [29] have used a PSO-decision tree technique and achieved 92.94% 5-fold cross-validation accuracy on SRBCT data set. A comparison of the different approaches on the MLL data set is shown in Table 5.8.

K. Yang et al. [20] employed the gene scoring algorithm, F-test strategies, and SVM and *k*-NN classifier to the SRBCT, ALL_AML, and MLL data sets. By their suggested method, there were too many informative genes that were chosen. M. S. Mohamad et al. [3] used an improved binary PSO for the same purpose on 10 distinct data sets, including the SRBCT, ALL_AML, and MLL data sets. For the SRBCT, ALL_AML, and MLL data sets, only 6, 2, and 4 genes, respectively, were shown to have 100% LOOCV accuracy.

The best subset of MLL data set is shown in Table 5.9. It is observed that 34306_at is highly expressed for MLL and it was also found by A. Sharma et al. [22]. A search in GEO profile also showed that it is moderate to highly expressed for MLL (GPL8300,34306_at (ID_REF),GDS2729). It has been also observed that 37710_at is highly expressed for ALL and for a few cases of MLL. This particular gene was also found by M. S. Mohamad et al. [3] and GEO profile showed that it is moderate to highly expressed for ALL and for a few cases of AML (GPL8300,37710_at (ID_REF),GDS1059). Again, 266_s_at is highly expressed for ALL and moderately expressed for a few cases of MLL. It was also found by M. S. Mohamad et al. [3] and GEO profile proves that it is highly expressed for ALL (GPL570,266_s_at (ID_REF),GDS2970). Moreover, 33054_at is found by the proposed method and it has been observed that it is moderately expressed for AML. A search in GEO profiles also showed a moderate expression for AML (GPL8300,33054_at (ID_REF),GDS1059) for this particular gene.

Finally, the selected subset of informative genes should have universal characteristics. That actually means the selected subset should be able to produce high classification accuracy on the test samples by applying a different classifier. Therefore, to assess the universal characteristic, classification accuracies are also evaluated using different kernels of SVM classifier. The classification accuracies on the three benchmark data sets are promising and are shown in Table 5.10. The obtained result reconfirms the usefulness and universal characteristics of the identified subset of informative genes. Further, the essential and sufficient conditions are used for a particular gene subset as a whole and hence there could be

Table 5.8 Comparison of the Methods on the MLL data set

Experiment	Method	Genes				Cross-Validation Accuracy (%)						Training Accuracy (%)	Test Accuracy (%)
K. Yang et al. [20]		5 CV		LOOCV		5 CV			LOOCV			–	–
		k-NN	SVM	*k*-NN	SVM	*k*-NN	SVM		*k*-NN		SVM		
	GS1	29	99	97	56	94.8	95.2		97.2		97.2	–	–
	GS2	91	87	90	91	94.9	94.7		97.2		97.2	–	–
	Chos	93	89	23	44	96	95.5		97.2		95.8	–	–
	F-test	99	100	65	31	95.4	94.8		95.8		95.8	–	–
A. Sharma et al. [22]	SFS + LDA with NCC	4				–						100	100
	SFS+ Bayes classifier	4				–						100	100
	SFS + NNC	4				–						100	93
M. S. Mohamad et al. [3]	IBPSO	4				100						–	–
B. Chandra et al. [28]	RNBC	–				10 CV						–	–
						RNBC		NBC		*k*-NN			
						87.14		80		68.57			
K. H. Chen et al. [25]	PSODT	–				5 CV						–	–
						100							
This work	Proposed technique	4				92.5439						100	100

Table 5.9 List of best subset of genes for MLL data

Index no. of selected genes	Gene Accession Number	Gene Name
841	33054_at	Hs.144852 gnl|*UG*|Hs#S1570396 Homo sapiens mRNA; cDNA DKFZp434N074 (from clone DKFZp434N074)
7930	34306_at	Hs.28578 gnl|*UG*|Hs#S892244 Homo sapiens KIAA0428 mRNA, complete cds
8937	37710_at	Hs.78995 gnl|*UG*|Hs#S2842 Homo sapiens MADS
12418	266_s_at	Hs.278667 gnl|*UG*|Hs#S885 Homo sapiens CD24 signal transducer mRNA, complete cds and 3' region

Table 5.10 Essential and sufficient conditions

Data sets	SVM Kernels	Training Accuracy (%)	Test Accuracy (%)
SRBCT	Linear	98.4127	95
	RBF with sigma = 1	98.4127	65
	Polynomial with order 3	100	90
	Quadratic	100	85
ALL_AML	Linear	97.3684	97.0588
	RBF with sigma = 1	100	97.0588
	Polynomial with order 3	100	94.1176
	Quadratic	100	97.0588
MLL	Linear	100	100
	RBF with sigma = 1	100	100
	Polynomial with order 3	100	100
	Quadratic	98.2456	93.3333

other such subsets. This demonstrates that the proposed method is sufficiently able to identify the cancer subgroups in terms of classification accuracy on blind test samples and number of informative genes. Therefore, biologists can directly refer to the selected subset of genes to diagnose cancer subgroups accurately.

5.5 Summary

In this chapter, a PSO-adaptive *k*-NN-based gene selection methodology is proposed to find a limited subset of useful genes for accurate categorization of cancer subgroups. A strategy for choosing the appropriate value of *k* for a *k*-NN classifier guided by PSO is also suggested. The dimension of the microarray data was previously reduced using several statistical techniques. Then only they have applied heuristic methods to select a small subset of informative genes. Therefore, the contribution of this work may be interpreted as the proposed gene selection methodology is capable of looking at all genes together and picking up whatever is needed. Additionally, by employing an SVM classifier, the usefulness and universality of the chosen subsets are proven. The findings of this study show that the suggested method is beneficial for identifying a limited fraction of discriminating genes that can distinguish between subgroups of microarray data related to cancer. Moreover, the proposed methodology is compared with other gene selection methods applied to the same data set and promising results are obtained by using the proposed methodology. It is observed that this method finds 6, 3, and 4 genes to achieve 100% (20/20), 97.0588% (33/34), and 100% (15/15) blind test accuracies for SRBCT, ALL_AML, and MLL data sets, respectively.

One of the most promising areas for using microarray technology is clinical biology. In order to evaluate several microbial genetic objects simultaneously, low- or middle-density microarrays are used. The suggested approach is centered on the potential for creating streamlined approaches for an effortless diagnosis of cancer subgroups by designing specialized microarray chips. As a result, histopathologists can use the chosen subset of relevant genes to help with the diagnosis and prognosis of various cancer subgroups. The advancement of microarray ancillary technologies must also be developed in order to fully realize the benefit of the technology for cancer subgroup detection based on microarrays. Together with statistics, software, and data mining technologies, new microarray platforms should emerge. In a broad sense, patients will gain from the accurate and less expensive technology.

For the purpose of creating meta-profiles in this area of cancer research, it is also necessary to develop standardized diagnostic methods that use biological imaging. The detection and grading of tumors in this context can be accomplished using cell level imaging data. The focus of the following chapter is on this aspect of cancer detection.

References

1. Shen Q, Shi W-M, Kong W (2008) Hybrid particle swarm optimization and tabu search approach for selecting genes for tumor classification using gene expression data. Comput Biol Chem 32:53
2. Lu Z, Xu Q, Al-Rajab M, Chiazor L (2021) Machine learning in cancer research with applications in colon cancer and big data analysis. IGI Global, Hershey
3. Mohamad MS, Omatu S, Deris S, Yoshioka M (2011) A modified binary particle swarm optimization for selecting the small subset of informative genes from gene expression data. IEEE Trans Inf Technol Biomed 15:813
4. Chen D, Liu Z, Ma X, Hua D (2005) Selecting genes by test statistics. J Biomed Biotechnol 2005:132
5. Eberhart R, Kennedy J (1995) MHS'95. In: Proceedings of the Sixth International Symposium on Micro Machine and Human Science. IEEE, pp 39–43
6. Vandeginste BM, Massart D, Buydens L, De Jong S, Lewi P, Verbeke J (1998) Handbook of chemometrics and qualimetrics, vol 20B. Elsevier
7. Dudoit S, Fridlyand J, Speed TP (2002) Comparison of discrimination methods for the classification of tumors using gene expression data. J Am Stat Assoc 97:77
8. Zhang M-L, Zhou Z-H (2007) ML-KNN: A lazy learning approach to multi-label learning. Pattern Recogn 40:2038

9. Tarlow D, Swersky K, Charlin L, Sutskever I, Zemel R (2013) International conference on machine learning. PMLR, pp 199–207
10. Ougiaroglou S, Nanopoulos A, Papadopoulos AN, Manolopoulos Y, Welzer-Druzovec T (2007) Advances in databases and information systems. In: 11th East European Conference, ADBIS 2007, Varna, 29 Sep–3 Oct 2007. Proceedings 11. Springer, pp 66–82
11. Kar S, Sharma KD, Maitra M (2015) Gene selection from microarray gene expression data for classification of cancer subgroups employing PSO and adaptive K-nearest neighborhood technique. Expert Syst Appl 42:612
12. Engelbrecht AP (2006) Fundamentals of computational swarm intelligence. Wiley, Hoboken
13. Khan J, Wei JS, Ringner M, Saal LH, Ladanyi M, Westermann F, Berthold F, Schwab M, Antonescu CR, Peterson C (2001) et al. Classification and diagnostic prediction of cancers using gene expression profiling and artificial neural networks. Nat Med 7:673
14. Tibshirani R, Hastie T, Narasimhan B, Chu G (2002) Diagnosis of multiple cancer types by shrunken centroids of gene expression. Proc Natl Acad Sci 99:6567
15. Lee C-P, Lin W-S, Chen Y-M, Kuo B-J (2011) Gene selection and sample classification on microarray data based on adaptive genetic algorithm/k-nearest neighbor method. Expert Syst Appl 38:4661
16. Pal NR, Aguan K, Sharma A, Amari S-I (2007) Discovering biomarkers from gene expression data for predicting cancer subgroups using neural networks and relational fuzzy clustering. BMC Bioinf 8:1
17. Fu LM, Fu-Liu CS (2005) Evaluation of gene importance in microarray data based upon probability of selection. BMC Bioinf 6:1
18. Bhattacharyya C, Grate L, Rizki A, Radisky D, Molina F, Jordan MI, Bissell MJ, Mian IS (2003) Simultaneous classification and relevant feature identification in high-dimensional spaces: application to molecular profiling data. Signal Process 83:729
19. Yeo G, Poggio T (2001) Technical Report, AI Memo 2001-018 CBCL Memo
20. Yang K, Cai Z, Li J, Lin G (2006) A stable gene selection in microarray data analysis. BMC Bioinf 7:1
21. Ji G, Yang Z, You W (2010) PLS-based gene selection and identification of tumor-specific genes. IEEE Trans Syst Man Cybern Part C (Appl Rev) 41:830
22. Sharma A, Imoto S, Miyano S (2011) A top-r feature selection algorithm for microarray gene expression data. IEEE/ACM Trans Comput Biol Bioinf 9:754
23. Zainuddin Z, Ong P (2011) Reliable multiclass cancer classification of microarray gene expression profiles using an improved wavelet neural network. Expert Syst Appl 38:13711
24. Li X, Shu L (2009) Kernel based nonlinear dimensionality reduction for microarray gene expression data analysis. Expert Syst Appl 36:7644
25. Chen K-H, Wang K-J, Tsai M-L, Wang K-M, Adrian AM, Cheng W-C, Yang T-S, Teng N-C, Tan K-P, Chang K-S (2014) Gene selection for cancer identification: a decision tree model empowered by particle swarm optimization algorithm. BMC Bioinf 15:1
26. Kumar PG, Victoire TAA, Renukadevi P, Devaraj D (2012) Design of fuzzy expert system for microarray data classification using a novel genetic swarm algorithm. Expert Syst Appl 39:1811
27. Wong T-T, Liu K-L (2010) A probabilistic mechanism based on clustering analysis and distance measure for subset gene selection. Expert Syst Appl 37:2144
28. Chandra B, Gupta M (2011) Robust approach for estimating probabilities in Naïve–Bayes Classifier for gene expression data. Expert Syst Appl 38:1293
29. Li H, Nantasenamat C, Monnor T, Isarankura-Na-Ayudhya C, Prachayasittikul V (2013) Genetic algorithm search space splicing particle swarm optimization as general-purpose optimize. Chemom Intell Lab Syst 128:153

Chapter 6
Intelligent Computing Approach for Leukocyte Identification

Abstract An automated approach for identifying and classifying leukocytes in microscopic blood smear images have been analyzed in this chapter. The approach involves identifying leukocytes using color-based clustering and then extracting a large set of features from the detected cells. Weighted aggregation-based transposition PSO (WATPSO) is introduced that effectively selects an optimized subset of features that are necessary for correctly classifying healthy cells and blast cells. An acute lymphoblastic leukemia (ALL) benchmark data set is utilized to test the suggested methodology. Remarkably, the result showcases a test accuracy of 97.30% with only 6 optimal features chosen by the WATPSO method. The experimental results highlight the effectiveness of this methodology, highlighting the use of few informative features and excellent test accuracy.

6.1 Introduction

Acute and chronic leukemia are the two main categories under which leukemia can be generically categorized. Compared to chronic leukemia, acute leukemia advances more quickly. Acute lymphoblastic leukemia (ALL) and acute myeloid leukemia (AML) are the two different subtypes of acute leukemia. The two age groups where ALL is most prevalent are children under 5 and adults over 50. It is a dangerous blood pathology that, if untreated, can be fatal. Therefore, a patient's life depends on receiving a correct diagnosis of ALL. Leukocyte analysis plays a significant role in the accuracy of the diagnostic prediction of ALL. Expert hematologists currently analyze leukocytes by looking at images of microscopic blood smears taken at the cell level. However, the accuracy of leukocyte detection and classification with this manual technique largely depends on the knowledge and skills of the hematologist. As a result, a sophisticated system for detecting and classifying leukocytes is actually needed as an additional predictive mechanism to help hematologists make their diagnostic predictions.

K. Das Sharma et al., *Intelligent Computing in Carcinogenic Disease Detection*, Computational Intelligence Methods and Applications,
https://doi.org/10.1007/978-981-97-2424-6_6

Numerous studies have focused on the identification and classification of leukocytes. Leukocytes have been detected using k-means clustering, shadowing C-means clustering, local binary pattern (LBP), and Hausdorff dimension (HD) [1–4]. Classification of healthy and blast leukocytes have been performed by analyzing different morphological features and color-based features [5, 6]. Researchers have proposed feature-based classification algorithms, but their classification accuracies are not very good [1–10]. The existence of redundant features utilized for classification purposes is one of the causes of that. Consequently, it is essential to choose a limited group of relevant features that can effectively distinguish between healthy and blast leukocytes. To identify pertinent features in this regard, many feature selection strategies have been proposed [3, 6, 11]. In this context, evolutionary optimization technique-based feature selection techniques have also been proposed by researchers [12–14].

In this chapter, the whole leukocyte has been segmented using a color-based k-means clustering technique on RGB color space. The roundness factor of the leukocytes have then been used to identify both individual and grouped leukocytes [10]. Grouped leukocytes have further been segmented into individual leukocytes by utilizing distance transform and watershed transform methodologies [15]. The "G" component of RGB color space is then subjected to the global thresholding approach to produce a binary mask. The mask thus obtained actually does take an active part to segment the cell nucleus. After that, the cytoplasm is extracted by executing an image subtraction operation between the nucleus and the whole leukocyte. The suggested method is necessary for a thorough analysis of every cell component, including the nucleus, cytoplasm, and whole leukocyte. To further categorize blast cells, a multitude of parameters are extracted, including shape-based, color-based, and textural information for the whole leukocyte, cell nucleus, and cytoplasm. Large-scale feature analysis will, however, require more processing power. Once more, an excessive amount of features include redundant features, which worsen the classification performance. Hence, for increasing the performance of classification accuracy, it is crucial to choose pertinent discriminative features from the vast array of input features.

In this chapter, a weighted aggregation-based transposition PSO (WATPSO) is used to perform the relevant feature selection task which can produce high classification accuracy with a near-smallest optimal feature subset. The proposed WATPSO divides the main swarm into two sub-swarms, thereby allowing the sub-swarms to perform a heterogeneous search to achieve two different objectives in the entire search space. Here, we have logically reasoned that finding a small subset of effective features and increasing the classification accuracy should be the two different objectives. Consequently, each sub-swarm is compelled to perform exploration of dissimilar region of search space, associated with each objective, by introducing different weights. Moreover, to converge to an optimum solution, the sub-swarms have been endorsed to share their information amongst themselves so as to ensure that both the objectives are satisfied at the best [16].

To achieve the goal, first, *k* nearest neighbor (*k*-NN) classifier with dynamic input feature space is used in this work for classification of blast cells. However, for *k*-NN-based classification problem, the value of *k* has a great influence on classification accuracy. If the value of *k* is too less the classifier lacks the essential robustness. On the contrary, if the value of *k* is too high the classifier faces poor classification accuracy. Therefore, proper selection of k is required [17] that would pilot the search into appropriate neighborhood and influence the classification accuracy for the better result.

This chapter focuses on finding non-redundant features using only training samples and then achieving high classification accuracy on blind test samples. The fitness function of the proposed WATPSO technique is constructed in such a way that higher classification accuracy is achieved with the help of a small subset of non-redundant features. Therefore, the proposed method can be utilized as a reliable decision support system for hematologists for accurate and early diagnosis of ALL.

6.2 Preprocessing

A preprocessing stage is required to prepare the sample microscopic blood smear images for classification. In this chapter, leukocytes have been identified first from the RGB image using color-based clustering technique. Then grouped leukocytes and individual leukocytes have been identified and separated using the roundness value of them. Grouped leukocytes have further been segmented into individual leukocytes by applying distance transform and watershed transform techniques. Thereafter, whole individual leukocytes have been recognized by removing abnormal components from the image depending on their solidity measures. Nucleus and cytoplasm have been then segmented from the whole leukocyte. A large number of features have been extracted from each cell elements such as nucleus and cytoplasm and also from the whole leukocyte. The WATPSO technique has further been employed for solving two objectives: finding a small subset of non-redundant features from the large feature set and increasing the performance of classification accuracy. The proposed methodology is shown in Fig. 6.1.

6.2.1 *Identification of Leukocytes*

Extracting the leukocytes from the blood smear images is the goal of leukocyte identification. The efficiency of a feature-based classification system depends on the precise segmentation of leukocytes. In this chapter, segmentation of leukocytes has been performed using color-based clustering in RGB color space. In the color-based clustering, the pixels with "R," "G," and "B" values have been treated as data points. Each data point has been then assigned to the closest centroid and each group of points assigned to a centroid is a cluster. Then the centroid of each cluster has

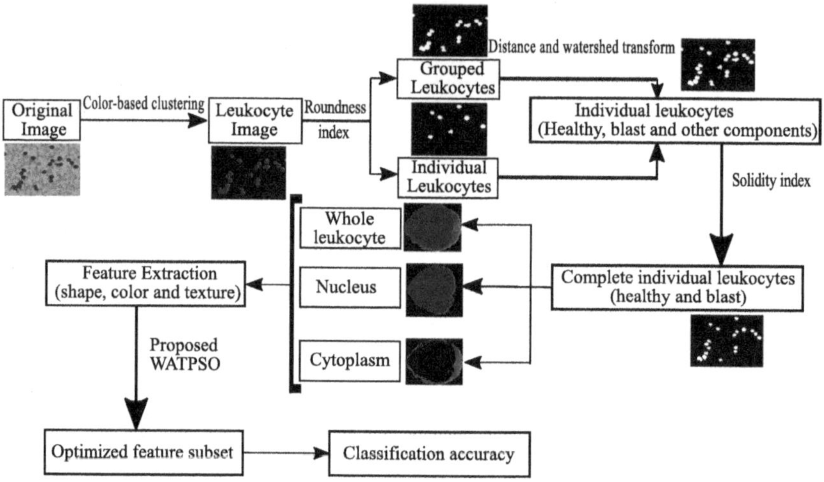

Fig. 6.1 Block diagram of the methodology for optimized feature selection for identification of carcinogenic leukocytes

been updated based on the points assigned to the cluster. The steps are repeated until no point changes cluster or until the clusters remain the same. However, to assign points to the closest centroid, a proximity measure is needed that quantifies the notion of "closes" for the specific data points under consideration. Here, Euclidean distance has been used as a distance measure for assigning data points to the closest centroid. Therefore, the objective of the color-based clustering is to minimize the intra-cluster variance or squared error function:

$$J = \sum_{j=1}^{c} \sum_{i=1}^{n} \left\| x_i - c_j \right\|^2 \tag{6.1}$$

where $\left\| x_i - c_j \right\|$ is the Euclidean distance between data point x_i and cluster-center c_j, n is the number of data points and c is the number of cluster centers. The leukocyte identification procedure using color-based clustering technique is elaborated in Algorithm 6.1 and is shown in Fig. 6.2.

Algorithm 6.1 Identification of leukocytes

```
for each RGB image do
    perform color-based clustering
    level every pixel in the image by cluster index
    separate the cluster containing leukocytes using mean cluster-center value
end for
```

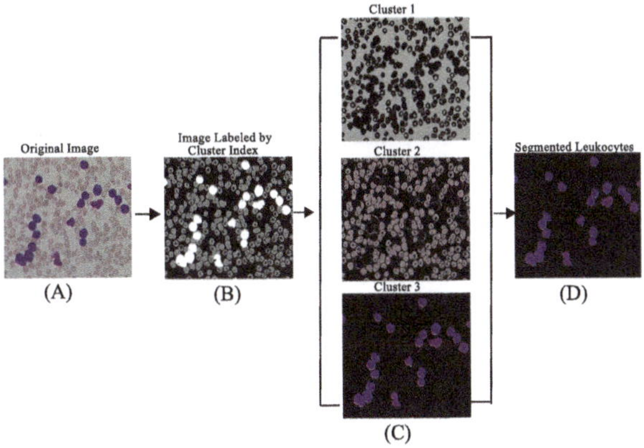

Fig. 6.2 Identification of leukocytes. (**a**) Original blood smear image (**b**) Image labeled by cluster index (**c**) Cluster1, cluster2 and cluster3 (**d**) Segmented leukocytes

6.2.2 Separation of Grouped Leukocytes

The procedures of feature extraction, feature selection, and classification can be greatly impacted by the grouped leukocytes as illustrated in Fig. 6.2. Therefore, identification and separation of grouped leukocytes are crucial. Many methodologies have been reported toward the detection and separation of grouped leukocytes [15]. In this chapter, the roundness value of leukocytes has been employed to discriminate grouped and individual leukocytes. Roundness is a measure of circularity that excludes the local irregularities [10]. Roundness value 1 can be used to represent a circular object, while roundness values less than 1 can be used to represent all other shapes. Thus, a threshold value can distinguish between leukocytes that are grouped together and those that are not. In this chapter, roundness value of 0.8 has been empirically selected as a threshold value. Hence, the objects whose roundness values are greater than or equal to the predefined threshold value have been identified as individual leukocytes and shown by red colored "∗" in Fig. 6.3b. The objects with roundness values less than the threshold are, thereby, identified as grouped leukocytes and also shown by blue colored "×" in Fig. 6.3b. The process of

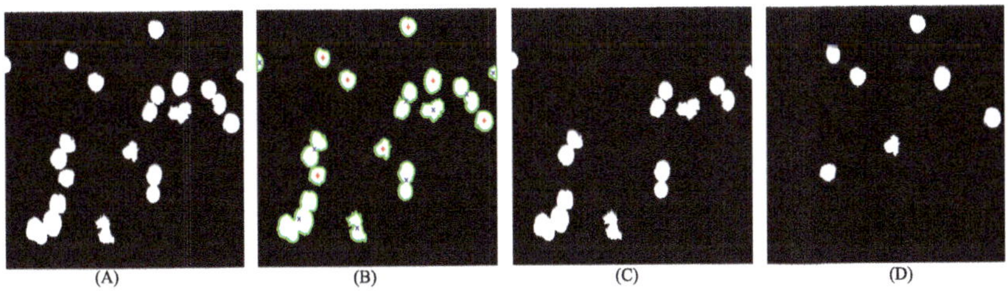

Fig. 6.3 Identification of grouped and individual leukocytes. (**a**) Binary leukocyte image (**b**) Image labeled by roundness index (**c**) Grouped leukocytes (**d**) Individual leukocytes

Algorithm 6.2 Identification of grouped and individual leukocytes

```
for each leukocyte do
    calculate the roundness value
    if roundness≥ predefined value then
        mark the leukocyte as individual leukocyte
    else
        mark the leukocyte as grouped leukocyte
    end if
end for
```

Algorithm 6.3 Separation of grouped leukocytes

```
for each grouped leukocyte do
    perform distance transform
    perform watershed transform
    perform minima imposition
    repeat watershed transform
end for
```

separation of grouped and individual leukocytes is explained in Algorithm 6.2 and is shown in Fig. 6.3.

To segment the grouped leukocytes into individual leukocytes, separation of adjacent cells are necessary. Therefore, first, distance transform and then watershed transform have been employed [15]. The watershed transform is based on visualizing the image in three dimensions and has been defined on the grayscale image. The pixel positions in spatial coordinate have been represented by x and y axis and the intensity of each pixel has been represented by z axis. In this work, we are only interested in the deepest minima and not all the other minima caused by the background texture to separate the adjacent cells of grouped leukocytes. Therefore, minima imposition technique has been employed in this work to retain the deepest minima and to remove the other minima [15]. A marker image that specifically sets some pixels to the lowest values has been made in order to identify the minima of interest and provide the image that highlights just the deepest minima while excluding others. To eventually divide the grouped leukocytes, watershed segmentation has been performed again. The process of separating grouped leukocytes is shown in Algorithm 6.3 and in Fig. 6.4.

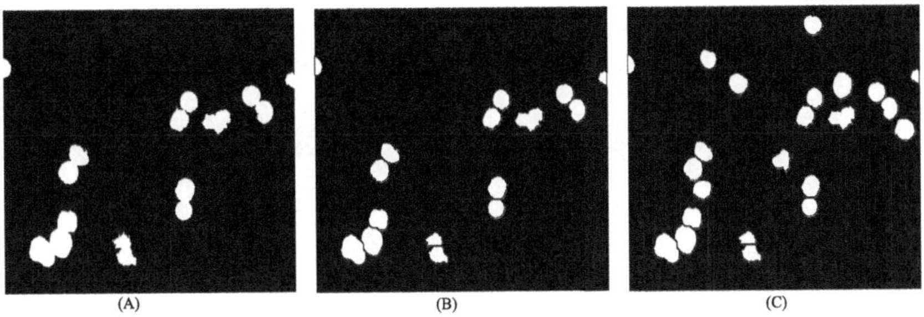

Fig. 6.4 Separation of grouped leukocytes. (**a**) Grouped leukocyte image (**b**) Separation of grouped leukocytes (**c**) Image including all detected leukocytes

6.2.3 *Image Cleaning*

Some border touching elements and other insignificant elements like unwanted dyes have been noticed in Fig. 6.4 after the separation of grouped leukocytes. As a result, as was previously indicated, all undesirable elements have been eliminated during the image cleaning stage. It is again necessary to eliminate unimportant components in order to increase classification accuracy. In particular, a straightforward border clearing technique can be used to easily remove the components that have merged with the boundaries [15]. In this chapter, the abnormal insignificant components have been detected by using the solidity measure, represented as:

$$S = \frac{A}{A_C} \tag{6.2}$$

where S denotes the solidity measure of the component, A denotes the area of the components, and A_C denotes the area of the convex hull of the components. Generally, the solidity index is a measure of the overall concavity of a component. As the area of the convex hull of the component approaches closer to the area of the component, the value of the solidity index tends to 1. As the component deviates from a closed circle the area of the convex hull increases resulting in the solidity index value to converge to less than 1. In this work, the solidity index value of 0.9 has been empirically selected as a threshold value. Hence, the objects whose solidity index values are greater than or equal to the predefined threshold value have been identified as individual leukocytes. The objects with solidity index values less than the threshold have been, thereby, identified as abnormal insignificant components and shown by blue colored "×" in Fig. 6.5a. The process of removing insignificant components from an image is shown in Algorithm 6.4. The image with abnormal insignificant components and final leukocyte image are shown in Fig. 6.5.

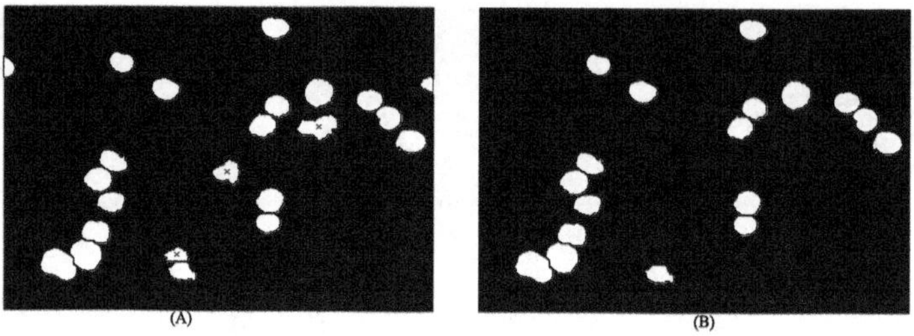

Fig. 6.5 Final separation leukocytes. (**a**) with abnormal insignificant components (**b**) Final leukocyte image

Algorithm 6.4 Image cleaning

for each leukocyte **do**
  calculate the solidity index value
  if solidity≥ predefined value **then**
    mark the leukocyte as individual leukocyte
  else
    mark the leukocyte as abnormal insignificant component
  end if
end for

6.2.4 *Separation of Whole Leukocyte, Nucleus, and Cytoplasm*

The next step is to separate the cytoplasm and nucleus of the leukocytes once they have been identified. Using color-based segmentation in RGB color space, the nucleus and cytoplasm of the whole leukocyte are separated in this work. A cluster index matrix has been created following the RGB image's color-based clustering process. The cluster index for every pixel is specified in the cluster index matrix. The mean cluster-center values have subsequently been computed. After that, the values of the cluster centers are arranged in ascending order. The position of the sorted cluster-center indexes in the cluster index matrix is then used to generate the nucleus index matrix. After that, the pixels' brightness values are taken out of the cluster containing the leukocyte component. The pixels that contain the nucleus have then been identified by performing the global threshold operation [15]. Finally, the cytoplasm has been extracted by performing an image subtraction operation between the whole leukocyte and nucleus. The process of the separation of whole leukocyte, nucleus, and cytoplasm has been elaborated in Algorithm 6.5 and is shown in Fig. 6.6.

Algorithm 6.5 Separation of whole leukocyte, nucleus, and cytoplasm

for each RGB leukocyte **do**
  perform the color-based clustering
  calculate the mean cluster values
  sort mean cluster values in ascending order and mark the index accordingly
  calculate nucleus index matrix by finding the position of the indexes in the cluster index matrix
  find the corresponding 'G' component value
  apply global thresholding to segment nucleus
  perform subtraction operation to segment cytoplasm
end for

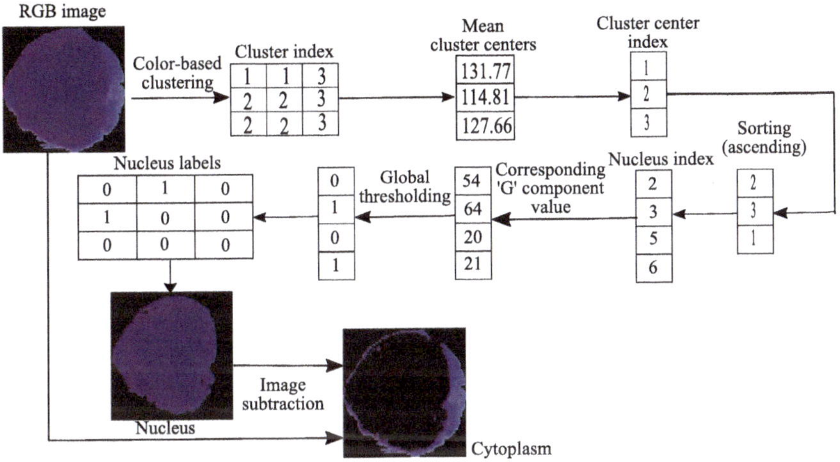

Fig. 6.6 Nucleus, cytoplasm and whole leukocyte separation process

6.3 Feature Extraction

Evaluating a set of features from the divided leukocytes, nucleuses, and cytoplasms is the goal of feature extraction. It is noteworthy that three distinct feature types – shape-based features [10], color-based features [10], and texture-based features [18] – have been recovered in this work from the nucleus, cytoplasm, and entire leukocytes. Utilizing the binary image of the nucleus, cytoplasm, and whole leukocyte, shape features have been retrieved. The 18 shape features [10] that are used in this work are listed in Table 6.1.

Furthermore, the shape features are susceptible to errors in segmentation. Therefore, these features have been used together with some color-based features [10]. These color-based features have been calculated from the grayscale images. There are six color-based features used in this work and have been described in Table 6.2.

Table 6.1 Shape-based features

Index no.	Feature name	Description
1	Area	Total number of pixels in the region
2	Convex area	Number of pixel in the convex hull of the region
3	Eccentricity	$\frac{\sqrt{(major\ \ axis^2 - minor\ \ axis^2)}}{major\ \ axis}$
4	Equivalent diameter	Diameter of a circle with the same area as the region
5	Euler number	Number of contiguous parts minus number of holes on a shape
6	Extent	Ratio of number of pixels in the region to the number of pixels in the bounding box
7	Filled area	Number of pixels in the region same size as the bounding box of the region
8	Major axis length	Number of pixel along the major axis of the ellipse that has the same normalized second central moment
9	Minor axis length	Number of pixel along the minor axis of the ellipse that has the same normalized second central moment
10	Orientation	Angle between x-axis and the major axis of the ellipse that has the same normalized second central moment
11	Perimeter	Distance along the boundary of the region
12	Solidity	Defined in Eq. (6.2)
13	Convex perimeter	Distance along the boundary of the convex hull of the region
14	Roundness	Measure of circularity that excludes the local irregularities
15	Elongation	$1 - \frac{minor\ \ axis}{major\ \ axis}$
16	Rectangularity	$\frac{area}{major\ \ axis \times minor\ \ axis}$
17	Compactness	$\frac{4 \times \pi \times area}{perimeter^2}$
18	Convexity	$\frac{Convex\ \ perimeter}{perimeter}$

Table 6.2 Color-based features

Index no.	Feature name	Description
1	Max intensity	Maximum intensity value of the region
2	Min intensity	Minimum intensity value of the region
3	Mean intensity	Mean intensity value of the region
4	Standard deviation	Standard deviation among the pixel values of the region
5	Skewness	$\frac{\sum_{i=1}^{n}(x_i-\mu)^3/n}{\sigma^3}$
6	Kurtosys	$\frac{\sum_{i=1}^{n}(x_i-\mu)^4/n}{\sigma^4}$

n is the total number of data points (pixels), μ is the mean value of the region and σ is the standard deviation of the region

These color-based features provide the information regarding the intensity distribution of the whole leukocyte, nucleus, and cytoplasm.

However, the location of the pixels with the same grey level should also be taken into account in addition to the intensity distribution [18]. The grey level co-occurrence matrix (GLCM) serves as the foundation for these properties. These features usually carry the texture information and are calculated from GLCM. The GLCM can be formed according to the Algorithm 6.6. The GLCM is a square matrix with dimension $M_g \times M_g$, where M_g is the number of gray levels in the image. Each element (i, j), of the matrix G has been constructed by calculating the number of times a pixel of intensity i is adjacent to a pixel with intensity j and then dividing the entire matrix by the total number of such comparisons. Each element of G has been then considered as the probability of the event that pixel i is adjacent with the pixel j. The GLCM can be expressed as Eq. (6.3). Haralick et al. [18] have derived features from the GLCM matrix which are illustrated below in Table 6.3. The adjacency of the pixels in Eq. 6.3 can be judged for angles 0°, 45°, 90° and 135° as shown in Fig. 6.7. Consequently, GLCM features enumerated in Table 6.3 have been evaluated for the four angles resulting in a total of 80 GLCM feature.

$$G = \begin{bmatrix} p(1,1) & p(1,2) & \ldots & p\left(1, M_g\right) \\ p(2,1) & p(2,2) & \ldots & p\left(2, M_g\right) \\ \ldots & \ldots & \ldots & \ldots \\ p\left(M_g, 1\right) & p\left(M_g, 2\right) & \ldots & p\left(M_g, M_g\right) \end{bmatrix} \tag{6.3}$$

Furthermore, the nucleus, cytoplasm, and whole leukocytes are used to calculate the 80 GLCM features in Table 6.3, 6 color features in Table 6.2, and 18 shape features in Table 6.1. Consequently, 312 features in total were taken into consideration for this chapter.

Algorithm 6.6 GLCM algorithm

let s be the pixel under consideration for calculation
let w be the set of pixels surrounding pixel s which will fall within a window centered upon pixel s
for the pixels in the set w **do**
  define each element of the GLCM as the number of times two pixels of intensity i and j occur in specified spatial relationship as shown in Fig. 6.7
end for
divide each element by the sum of all elements

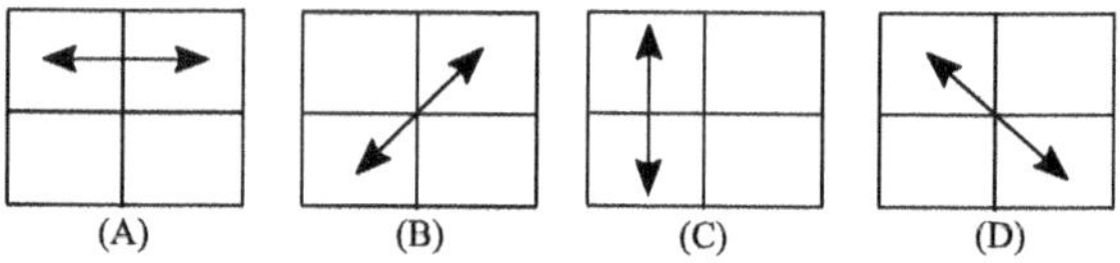

Fig. 6.7 Four directions of adjacency as defined for calculation of the GLCM texture features (**a**) 0° (**b**) 45° (**c**) 90° (**d**) 135°

Thus, an extensive set of 312 features has been developed following a thorough preprocessing, and this feature set is now undergoing a WATPSO-based optimal feature selection procedure.

6.4 Weighted Aggregation-Based Transposition PSO for Feature Evaluation

As previously noted, there must be a significant computational cost because to the enormous number of features. Additionally, there is a chance that some of the features in the set are redundant, which could decrease classification accuracy in addition to growing the size of the feature subset. Therefore, in order to attain improved classification accuracy at a computationally affordable cost, it is required to identify the subset of characteristics that may function as a set of optimal features. We stress the validity of our suggested WATPSO method in this sense, as it is effective enough to choose the best feature subset and get high classification accuracy with a minimal number of differentiating features.

First, in the suggested method, the WATPSO receives all of the retrieved features as input. After that, the WATPSO approach is used to analyze several candidate feature sets. The k-NN classifier uses the candidate feature sets, which vary in amount of features, to create a dynamic input feature space. The k-NN classifier's output and dynamic input feature space are then used to evaluate the fitness function. The performance of the classification of the blind test samples is then examined by extracting an optimized feature space from the dynamic input feature space. The proposed WATPSO-based optimized feature subset evaluation process is shown in Fig. 6.8 where the training and testing phases have been shown in solid line and dashed line, respectively.

Table 6.3 GLCM features

Index no.	Feature name	Description
1	Energy	$\sum_i \sum_j p(i,j)^2$
2	Contrast	$\sum_{n=0}^{M_g-1} n^2 \left\{\sum_{i=1}^{M_g} \sum_{j=1}^{M_g} p(i,j)\right\}$, $\|i-j\| = n$
3	Correlation	$\frac{\sum_i \sum_j (ij) p(i,j) - \mu_x \mu_y}{\sigma_x \sigma_y}$
4	Sum of squares (variance)	$\sum_i \sum_j (i-\mu)^2 p(i,j)$
5	Inverse difference moment	$\sum_i \sum_j \frac{1}{1+(i-j)^2} p(i,j)$
6	Sum average	$\sum_{i=2}^{2M_g} i p_{x+y}(i)$
7	Sum entropy	$-\sum_{i=2}^{2M_g} p_{x+y}(i) log\left\{p_{x+y}(i)\right\} = f_8$
8	Sum variance	$\sum_{i=2}^{2M_g} (i-f_8)^2 p_{x+y}(i)$
9	Entropy	$-\sum_i \sum_j p(i,j) log(p(i,j))$
10	Difference variance	$\sum_{i=0}^{M_g-1} i^2 p_{x-y}(i)$
11	Difference entropy	$-\sum_{i=0}^{M_g-1} p_{x-y}(i) log(p_{x-y}(i))$
12	Information measure of correlation 1	$\frac{HXY-HXY1}{max\{HX,HY\}}$
13	Information measure of correlation 2	$(1-exp[-2(HXY2-HXY)])^{\frac{1}{2}}$
14	Autocorrelation	$\sum_i \sum_j (ij) p(i,j)$
15	Dissimilarity	$\sum_i \sum_j \|i-j\| p(i,j)$
16	Homogeneity	$\sum_i \sum_j \frac{p(i,j)}{1+(i-j)^2}$
17	Inverse difference	$\sum_i \sum_j \frac{p(i,j)}{1+\|i-j\|}$
18	Maximum probability	$max_{i,j} p(i,j)$
19	Cluster prominence	$\sum_i \sum_j (i+j-2\mu)^3 p(i,j)$
20	Cluster shed	$\sum_i \sum_j (i+j-2\mu)^4 p(i,j)$

μ_x, μ_y, σ_x and σ_y are the means and standard deviation of p_x and p_y, the probability density function, x and y are the co-ordinates of an element in G, $p_{x+y}(i)$ is the probability of co-occurrence matrix co-ordinates summing to $x+y$, HX, HY are the entropies of p_x and p_y, $HXY = -\sum_i \sum_j p(i,j) log(p(i,j))$, $HXY1 = -\sum_i \sum_j p(i,j) log\left\{p_x(i) p_y(j)\right\}$, $HXY2 = -\sum_i \sum_j p_x(i) p_y(j) log\left\{p_x(i) p_y(j)\right\}$

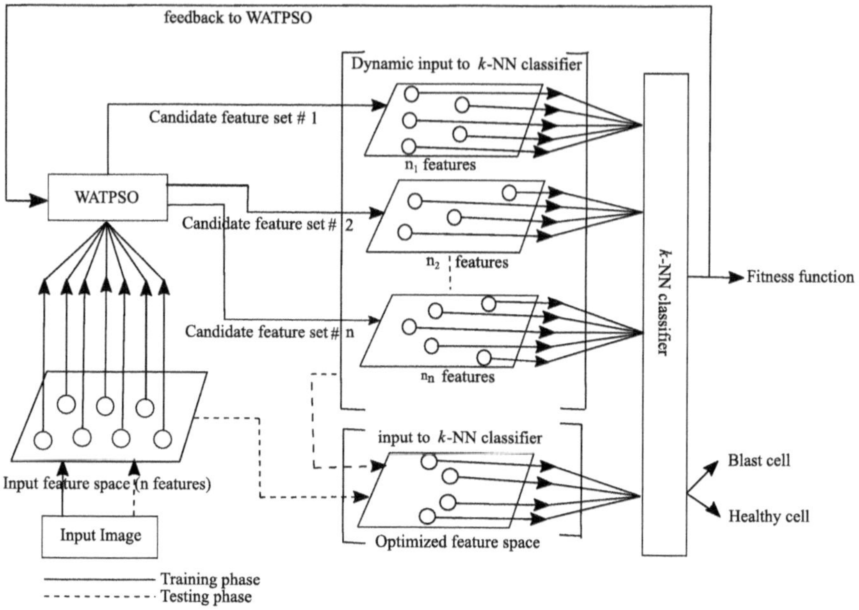

Fig. 6.8 Proposed WATPSO-based optimized feature subset evaluation process

6.4.1 *WATPSO-Based Feature Selection Technique*

The PSO algorithm is a proficient method for identifying the best solutions for continuous nonlinear functions. In conventional PSO [19], each particle's position and velocity are updated in real time by monitoring both its individual best information and the global best information of the entire swarm. Wang et al. [20] have presented a variation of classical PSO termed as binary PSO with mutation and time-varying acceleration coefficients. By injecting random mutation during the position update stage, the mutation operator aids in improving the particle position. At the beginning of the search, the time-varying acceleration coefficients improve the global search capability; later on, each particle's local search capability is improved. Another variant of PSO, namely predator-prey PSO, has been proposed by Zhang et al. [21]. The whole swarm in the predator-prey PSO is split into two sub-swarms: the predator sub-swarm and the prey sub-swarm. The main concept is that, in order to improve the optimization efficiency of the traditional PSO, the predator sub-swarm chases the prey sub-swarm and the prey sub-swarm attempts to escape from the predator sub-swarm. Nonetheless, the optimization procedure in this study must address a number of goals, two in particular. The commonly used techniques to handle multiple objectives are to use separate fitness functions for each objective and to employ Pareto optimal solution of the process, if any. Two approaches can be considered which might lead to an acceptable solution. The first consideration could be searching for an algorithm through which one cannot obtain a single solution at the end, rather get a set of all Pareto optimal solutions in the entire search space. Moreover, the designer has to make a tedious trade-off between the Pareto dominance and crowding of probable solution points at the Pareto front [22].

The second consideration toward the solution is weighted aggregation-based approach [23]. In this approach, the solution could be obtained by systematically changing the weights during evolution. We adhere to the second approach. And hence, in weighted aggregation, the fitness function is defined by:

$$FF = \sum_{i=1}^{n} \omega_i . F_i \tag{6.4}$$

where w_i are non-negative weights such that

$$\sum_{i=1}^{n} \omega_i = 1 \tag{6.5}$$

As we discussed earlier, the objectives in this work are two-folds. One objective is to increase the classification accuracy and other is to reduce the subset size containing effective non-redundant features. Therefore, a novel WATPSO technique has been proposed so as to achieve the dual objectives. We, here take the liberty to propose WATPSO, an indigenous variant of typical PSO that divides the main swarm into two sub-swarms which can perform a heterogeneous search for each objective. The heterogeneous search has been performed by asserting different weights to different objectives. Two sub-swarms also share information with each other. The information sharing model is shown in Fig. 6.9. To perform the above-mentioned optimization of a dual objective function by WATPSO technique the fitness function in this work

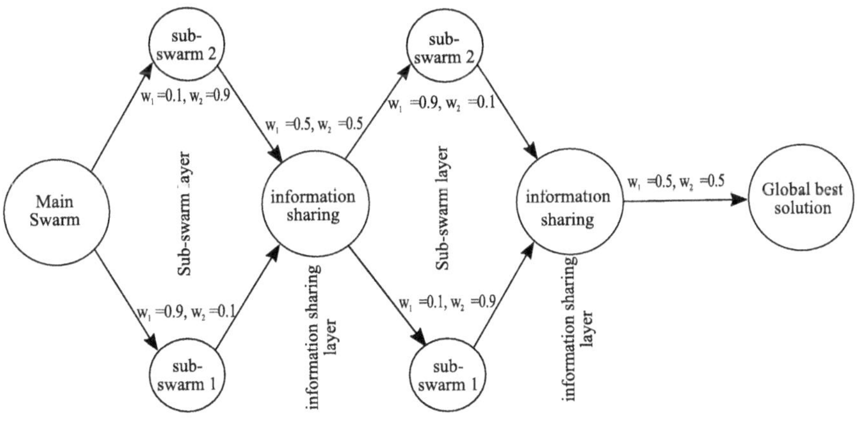

Fig. 6.9 Proposed WATPSO technique

is constructed by the following equation:

$$FF = \omega_1 \times \text{classification error} + \frac{\text{selected features}}{\text{total features}} \times \omega_2 \tag{6.6}$$

where $\omega_1 = (0.1, 0.9)$ and $\omega_2 = (1 - \omega_1)$ and classification error = error over classification accuracy.

In the first sub-swarm layer, the search process has been guided more toward minimizing the classification error by one sub-swarm by assigning a weight of 0.9 with the classification error objective. On the other hand, another sub-swarm guides the search process toward minimizing effective non-redundant features by assigning 0.9 weight with it in the same layer. After performing the first layer of search, the information of each sub-swarm has been shared by them with each other in the first information sharing layer. In the information sharing layer, two sub-swarms have been combined into one swarm and equal weights are assigned to them, i.e., ($\omega_1 = \omega_2 = 0.5$). The aim of the information sharing layer is to provide scope for each particle in each sub-swarm to share their information in the combined swarm. Then the heterogeneous search is again initiated by the two sub-swarms by altering their corresponding weights in the second sub-swarm layer. Thereafter another information sharing layer has been provided to share the particle information further. In the sub-swarm layer, the position and velocity of each particle are initialized between 0 and 1 as:

$$x_{i,j}^{ssl}(\Delta t) = S_{min} + r_1 (S_{max} - S_{min}) \tag{6.7}$$

$$v_{i,j}^{ssl}(\Delta t) = S_{min} + r_2 (S_{max} - S_{min}) \tag{6.8}$$

where ssl denotes sub-swarm layer, r_1 and r_2 are random numbers, S_{min} and S_{max} are the vectors of lower and upper bound, i.e., vector of 0's and 1's in the present case, Δt is the time step. In this work, the dimension of each particle has been assigned as equal to the dimension of the extracted features. Then, a threshold ϵ (selected experimentally by trial and error method as 0.9) has been utilized to construct the feature selection flag of binary bits using the following equation:

$$S_{feature} = \begin{cases} 1, & \text{for } S_{feature} \geq \epsilon \\ 0, & \text{otherwise.} \end{cases} \tag{6.9}$$

Therefore, each particle stands for a subset of features where "1" represents the effectiveness of the corresponding feature. These effective features have then been used as a candidate feature set to evaluate the fitness function. This candidate feature set selection process has been portrayed in Fig. 6.10. The personal best and global

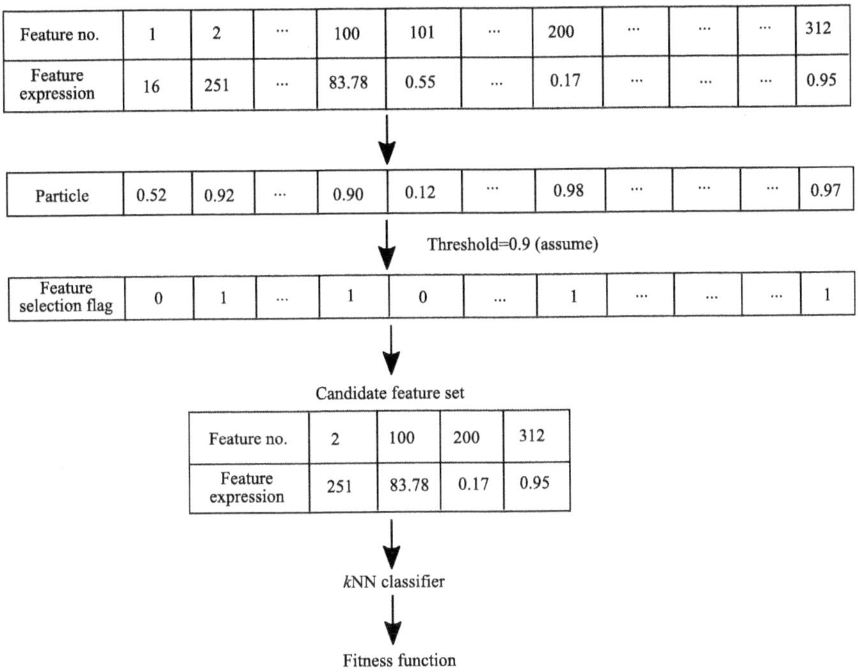

Fig. 6.10 Candidate feature set selection process

best solutions for each particle in the sub-swarm layer have been calculated as:

$$p_{i,j}^{ssl}(\Delta t+1)=\begin{cases}p_{i,j}^{ssl}(\Delta t), & \text{if } FF\left(x_{i,j}^{ssl}(\Delta t+1)\right)\geq FF\left(p_{i,j}^{ssl}(\Delta t)\right)\\ x_{i,j}^{ssl}(\Delta t+1), & \text{otherwise.}\end{cases} \tag{6.10}$$

$$g_{best}^{ssl}(\Delta t)\in min\{FF(p_0(\Delta t)), FF(p_1(\Delta t)), \ldots, FF(p_{ss}(\Delta t))\} \tag{6.11}$$

where $x_{i,j}^{ssl}$ is the current position of the particle in the sub-swarm layer, FF is the objective function, and ss is the size of the sub-swarm. The position and velocity of each particle in the sub-swarm layer have been updated using the following equations:

$$v_{i,j}^{ssl}(\Delta t+1)=\omega v_{i,j}^{ssl}(\Delta t)+c_1r_1\left[p_{i,j}^{ssl}(\Delta t)-x_{i,j}^{ssl}(\Delta t)\right] + c_2r_2\left[g_{best,j}^{ssl}(\Delta t)-x_{i,j}^{ssl}(\Delta t)\right] \tag{6.12}$$

$$x_{i,j}^{ssl}(\Delta t+1)=x_{i,j}^{ssl}(\Delta t)+v_{i,j}^{ssl}(\Delta t+1) \tag{6.13}$$

where ω represents the inertia weight, c_1, c_2 are acceleration constants, and r_1, r_2 are random numbers. The key concept of the proposed process is the transposing behavior of the weights associated with each objective in each layer, wherein, this transposing attribute of the proposed algorithm allows altering the weights of the sub-swarms dynamically. Moreover, as said earlier, the proposed transposition nature of sub-swarms has been devised so that the sub-swarms can explore different regions of the search space and ensures sharing of information with each other to enrich the information base of the main swarm. Indeed, the complete swarm in the information exchange layer is formed with particles having initial position and velocity as:

$$x_{i,j}^{iel}(\Delta t) = x_{i,j}^{ssl}(\Delta t + 1) \tag{6.14}$$

$$v_{i,j}^{iel}(\Delta t) = v_{i,j}^{ssl}(\Delta t + 1) \tag{6.15}$$

where iel represents information exchange layer. The position and velocity of each particle during information exchange have been updated as:

$$\begin{aligned} v_{i,j}^{iel}(\Delta t + 1) = \omega v_{i,j}^{iel}(\Delta t) + c_1 r_3 \left[p_{i,j}^{iel}(\Delta t) - x_{i,j}^{iel}(\Delta t) \right] \\ + c_2 r_4 \left[g_{best,j}^{iel}(\Delta t) - x_{i,j}^{iel}(\Delta t) \right] \end{aligned} \tag{6.16}$$

$$x_{i,j}^{iel}(\Delta t + 1) = x_{i,j}^{iel}(\Delta t) + v_{i,j}^{iel}(\Delta t + 1) \tag{6.17}$$

In essence, the dynamic weight based transposition, inspired by the crossover phenomenon of GA, makes the search process more robust toward satisfying the goal of achieving high classification accuracy utilizing less number of effective features. The flowchart of the proposed WATPSO methodology has been shown in Fig. 6.11. The dynamic transposition nature of weights associated with each objective is depicted in Table 6.4.

6.5 Experimental Study

6.5.1 *Database*

To evaluate the performance of the proposed approach, the acute lymphoblastic leukemia image database ALL-IDB [24] have been utilized. This is a public domain image database of peripheral blood samples from leukemia patients and healthy individuals. The database consists of two versions; ALL-IDB1 data set consists of total 108 images and can be used to verify the segmentation capability of the algorithms, whereas, ALL-IDB2 data set consists of cropped areas of normal and blast cells that belong to the images in ALL-IDB1 data set. The ALL-IDB2 data set consists of total 260 images. This ALL-IDB2 data set can only be used for testing the proposed feature selection and classification algorithm. Each image in ALL-IDB1

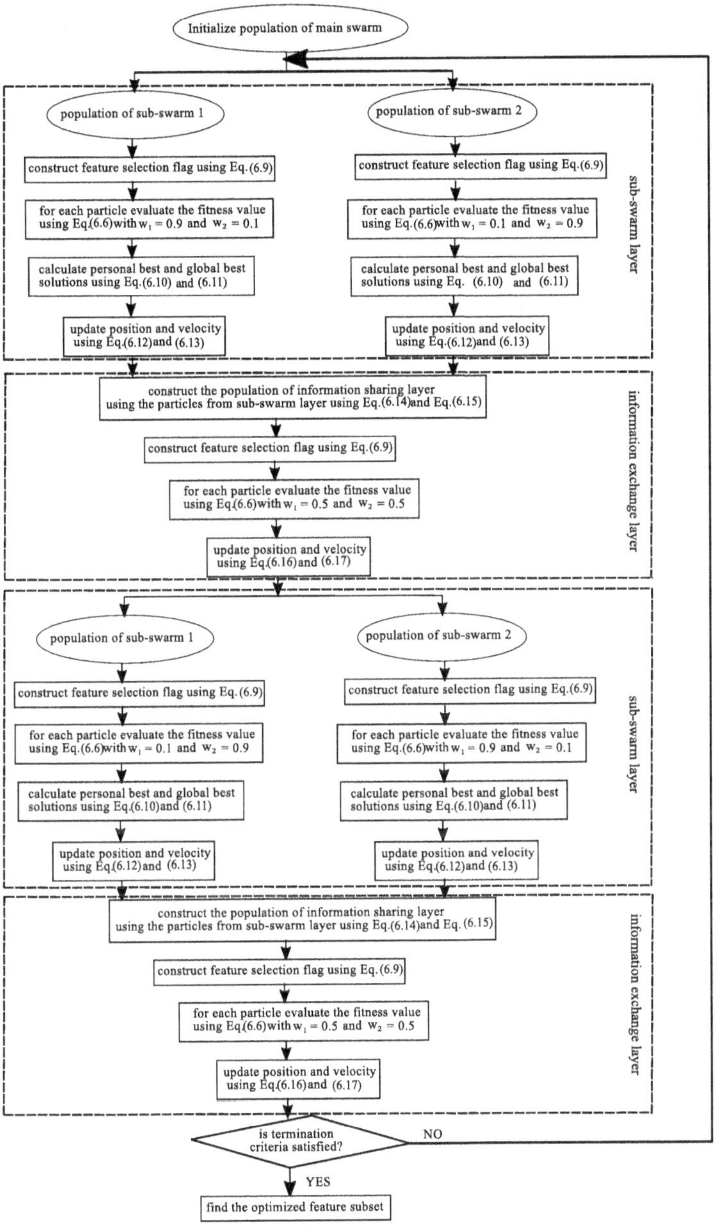

Fig. 6.11 Flowchart of the proposed WATPSO technique

Table 6.4 Weight selection for WATPSO

Layer	Sub-swarm1	Sub-swarm2	Recombined swarm	Fitness (sub-swarm 1)	Fitness (sub-swarm 2)	Fitness (recombined swarm)
Sub-swarm layer	$w_1 = 0.9$, $w_2 = 0.1$	$w_1 = 0.1$, $w_2 = 0.9$	–	accuracy = 93.10%, feature = 49	accuracy = 92.24%, feature = 24	–
Information exchange layer	–	–	$w_1 = 0.5$, $w_2 = 0.5$	–	–	accuracy = 88.79%, feature = 13
Sub-swarm layer	$w_1 = 0.1$, $w_2 = 0.9$	$w_1 = 0.9$, $w_2 = 0.1$	–	accuracy = 91.38%, feature = 10	accuracy = 92.67%, feature = 17	–
Information exchange layer	–	–	$w_1 = 0.5$, $w_2 = 0.5$	–	–	accuracy = 89.65%, feature = 6

and ALL-IDB2 consists of an associated text file carrying the centroid information for each blast cell.

6.5.2 Performance Measure

In the proposed approach, each particle has been allowed to represent a candidate solution vector depending upon the threshold ϵ. Each particle is responsible to evaluate the fitness function throughout the iteration. Moreover, different experiments have been independently conducted thirty times on the entire data set. To judge the performance of the proposed approach and also to compare with other approaches, some important criteria have been considered such as (a) sensitivity, (b) specificity and (c) accuracy. These criteria have been evaluated as:

$$Sensitivity = \left(\frac{TP}{TP + FN}\right) \times 100\% \tag{6.18}$$

$$Specificity = \left(\frac{TN}{TN + FP}\right) \times 100\% \tag{6.19}$$

$$Accuracy = \left(\frac{TP + TN}{TP + TN + FP + FN}\right) \times 100\% \tag{6.20}$$

where TP (true positive) represents the classification outcomes that are positive in case of blast cells, TN (true negative) represents the classification outcomes that are negative in case of healthy cells. Similarly, FP (false positive) represents the classification results that are positive in case of healthy cells and FN (false negative) represents the classification results that are negative for blast cells.

6.5.3 Performance Evaluation

In this work, the ALL IDB1 data set has been utilized to evaluate the performance of the proposed methodology. The performance of the proposed leukocyte identification technique has been shown in Table 6.5. The segmentation performance has been measured by Jaccard index (JI), dice coefficient (DC), volumetric similarity (VS), and reliability index (RI) [25]. The JI measures the similarity between the ground truth and the segmented object. The DC is used to measure the extent of spatial overlap between the ground truth and the segmented object and the VS measures the similarity between the ground truth and the segmented object by dividing the absolute volume difference by the sum of volumes. The range of JI, DC, and VS are between 0 and 1. If the segmented object and the ground truth are similar then

Table 6.5 Performance of leukocyte identification from ALL-IDB data set

Image	Annotated leukocytes	Leukocyte count by proposed method	Leukocyte identification accuracy	Jaccard Index (*JI*)	Dice Coefficient (*DC*)	Volumetric Similarity (*VS*)	Reliability Index (*RI*)
Image 1	8	6	75%	0.9621	0.9807	0.9335	0.9972
Image 2	8	8	100%	0.9985	0.9993	0.9993	0.9985
Image 3	11	11	100%	0.9511	0.9750	0.9771	0.9978
Image 4	6	6	100%	0.8991	0.9469	0.9480	0.9988
Image 5	23	17	73.91%	0.8652	0.9277	0.9304	0.9972
Image 6	15	14	93.33%	0.9134	0.9547	0.9559	0.9988
Image 7	5	5	100%	0.9992	0.9996	0.9996	0.9992
Image 8	15	14	93.33%	0.8984	0.9465	0.9482	0.9982
Image 9	6	6	100%	0.9202	0.9585	0.9586	0.9998
Image 10	11	10	90.90%	0.9204	0.9586	0.9592	0.9994
Image 11	13	12	92.30%	0.9993	0.9997	0.9997	0.9993
Image 12	11	11	100%	0.9990	0.9995	0.9995	0.9990
Image 13	10	6	60%	0.9992	0.9996	0.9996	0.9992
Image 14	3	3	100%	0.6084	0.7566	0.7569	0.9995
Image 15	15	15	100%	0.9981	0.9991	0.9991	0.9981
Image 16	15	15	100%	0.9560	0.9775	0.9779	0.9995
Image 17	2	2	100%	0.9982	0.9991	0.9991	0.9982
Image 18	5	5	100%	0.9569	0.9780	0.9785	0.9995
Image 19	11	11	100%	0.9208	0.9588	0.9612	0.9975

Image 20	1	1	100%	0.9994	0.9995	0.9997	0.9998
Image 21	2	2	100%	0.9995	0.9993	0.9992	0.9998
Image 22	2	2	100%	0.9979	0.9989	0.9989	0.9979
Image 23	5	5	100%	0.8680	0.9293	0.9293	0.9998
Image 24	3	3	100%	0.9991	0.9993	0.9995	0.9998
Image 25	2	2	100%	0.9992	0.9994	0.9996	0.9998
Image 26	4	4	100%	0.9991	0.9997	0.9996	0.9998
Image 27	1	1	100%	0.9995	0.9994	0.9993	0.9998
Image 28	1	1	100%	0.9997	0.9995	0.9994	0.9998
Image 29	2	2	100%	0.8540	0.9212	0.9212	0.9998
Image 30	1	1	100%	0.9994	0.9997	0.9997	0.9994
Image 31	1	1	100%	0.9993	0.9996	0.9994	0.9997
Image 32	1	1	100%	0.9995	0.9994	0.9996	0.9996
Image 33	1	1	100%	0.9955	0.9992	0.9992	0.9985
Image 48	16	16	100%	0.9374	0.9677	0.9688	0.9988
Image 49	13	13	100%	0.8778	0.9349	0.9355	0.9993
Image 50	12	12	100%	0.8099	0.8950	0.8959	0.9990

(continued)

Table 6.5 (continued)

Image	Annotated leukocytes	Leukocyte count by proposed method	Leukocyte identification accuracy	Jaccard Index (*JI*)	Dice Coefficient (*DC*)	Volumetric Similarity (*VS*)	Reliability Index (*RI*)
Image 51	17	16	94.11%	0.9617	0.9805	0.9830	0.9975
Image 52	23	20	86.95%	0.9676	0.9836	0.9863	0.9972
Image 53	32	31	96.87%	0.9638	0.9816	0.9825	0.9990
Image 54	10	10	100%	0.8158	0.8986	0.9009	0.9974
Image 55	20	19	95%	00.8364	0.9109	0.9128	0.9980
Image 56	12	9	75%	0.9466	0.9726	0.9746	0.9979
Image 57	11	11	100%	0.9984	0.9992	0.9992	0.9984
Image 58	21	21	100%	0.8304	0.9073	0.9084	0.9988
Image 59	30	30	100%	0.9977	0.9988	0.9988	0.9977
Image 60	11	11	100%	0.9156	0.9559	0.9573	0.9985
Image 61	21	17	80.95%	0.9981	0.9991	0.9993	0.9984
Image 62	20	20	100%	0.9069	0.9512	0.9542	0.9968
Image 63	20	16	80%	0.9991	0.9996	0.9996	0.9991
Average	10.38	9.71	95.67%	0.9436	0.9693	0.9690	0.9988

the values of JI, DC, and VS are close to 1.

$$JI = \frac{|S \cap GT|}{|S \cup GT|} \tag{6.21}$$

$$DC = \frac{2 \times |S \cap GT|}{|S| + |GT|} \tag{6.22}$$

$$VS = 1 - \left| \frac{|S| - |GT|}{|S| + |GT|} \right| \tag{6.23}$$

$$RI = \frac{|S \cap GT|}{|GT|} \tag{6.24}$$

where GT is the ground truth annotated by the expert radiologists and S is the segmentation result of the proposed method. The average leukocyte detection accuracy by the proposed method is found to be 95.67%.

Further, a feature matrix of size 582 × 312 has been extracted from the sub-images containing whole leukocyte, nucleus, and cytoplasm where total number of leukocytes (including blast and healthy cells) are 582 and total number of features are 312. As stated earlier, this 312 number of features may contain redundant features. These redundant features not only reduce the classification performance but also increase the computation cost. Therefore, the WATPSO-based feature selection technique has been utilized to select the optimized non-redundant features that can classify blast cells and healthy cells with high enough accuracy.

The proposed WATPSO technique has been applied to the 582 × 312 feature matrix to evaluate the optimized feature subset. This optimized feature subset has then been utilized to evaluate the classification accuracy on blind test samples. In the proposed work ALL-IDB2 data set has been used as blind test samples. The selection of effective feature subset and evaluation of blind test accuracy have been conducted thirty times. The computation time of the proposed methodology has also been calculated for thirty experiments. It has been shown that an average computation time of 771.09 s. is required to execute the proposed methodology in a hardware platform with Intel (R)Core (TM) i7-4510U CPU@2.00 GHz processor with 8.0 GB RAM. The results of thirty experiments have been shown in Table 6.6 where the best result achieved has been highlighted in gray shade. The nature of fitness function for the best-achieved result has also been shown in Fig. 6.12.

It is observed that 97.30% blind test accuracy has been achieved as a best result by using only 6 features. These optimal 6 features have actually been derived from whole leukocyte, nucleus, and cytoplasm and include shape-based features, color-based features and texture features. The description of the optimal feature set has been shown in Table 6.7.

Table 6.6 Result of ten independent runs of the proposed algorithm

Run	No. of features	Features	Sensitivity%	Specificity%	Training accuracy	Blind test accuracy%	Computation time (s)
1	14	9,16,44,66,75,119,136,144,164,192,273,274,276,290	91.11	85	89.34	89.23	911.07
2	11	16,53,89,102,120,146,159,224,260,274,282	94.44	81.66	91.40	81.15	903.77
3	6	39,51,119,191,224,283	96.96	86.66	99.65	81.15	848.23
4	6	84,118,166,186,195,270	91.89	85.45	92.09	86.53	800.32
5	6	8,18,44,119,224,283	94.59	86.36	90.03	91.53	797.94
6	9	18,39,90,107,114,143,166,201,224	94.11	80	87.80	87.69	775.74
7	8	8,23,44,94,119,143,166,191	97.14	86.66	92.09	91.92	777.31
8	7	53,69,159,166,216,224,273	96.55	87.14	91.06	88.84	772.21
9	6	18,44,88,102,159,273	96.55	87.17	91.40	89.23	847.83
10	7	44,51,119,135,217,224,226	98.95	83.33	97.42	97.30	746.47
11	8	88,91,141,152,191,242,273,285	96.66	86.66	92.09	91.92	747.89
12	4	44,137,171,192	96.66	86.66	92.43	91.92	831.32
13	5	119,143,164,224,252	96.42	80	89.34	86.15	809.25
14	6	85,101,166,191,242,273	96.66	86.66	92.43	91.92	708.36
15	7	18,159,175,181,183,224,254	96.55	81.42	94.67	94.61	704.43
16	6	51,119,143,158,232,284	96.42	80	86.76	86.53	734.33
17	13	18,44,65,119,135,158,159,166,199,224,258,274,279	98.95	81.42	95.01	95	739.21

18	4	30,70,166,279	96.29	84.44	83.84	83.46	713.80
19	6	26,64,119,161,270,273	98.95	83.33	97.76	97.30	748.92
20	8	36,44,51,119,191,224,262,273	96.55	87.14	89.34	88.84	758.97
21	6	18,44,119,143,191,224	96.66	86.66	92.09	91.53	743.50
22	11	44,80,119,158,166,179,224,258,273,282,310	98.95	81.42	95.01	94.61	736.55
23	6	16,44,88,159,166,189	96.55	87.14	91.40	89.23	734.36
24	8	36,44,51,119,191,224,262,273	96.66	86.66	92.43	91.92	744.86
25	12	37,44,88,114,119,133,135,199,217,235,239,264	98.95	81.42	95.36	95	732.65
26	12	44,51,65,59,119,143,191,217,224,273,277,283	96.66	86.66	92.78	92.30	732.23
27	9	18,34,51,81,119,131,158,224,252	96.66	86.66	92.43	91.92	756.62
28	6	36,119,143,152,224,273	96.42	80	87.80	86.53	733.39
29	7	75,119,142,224,258,276,297	98.95	83.33	97.76	97.30	813.14
30	5	35,39,44,179,266	93.33	80	91.06	86.15	727.01
Avg.	8	–	96.40	84.23	92.13	90.25	771.09

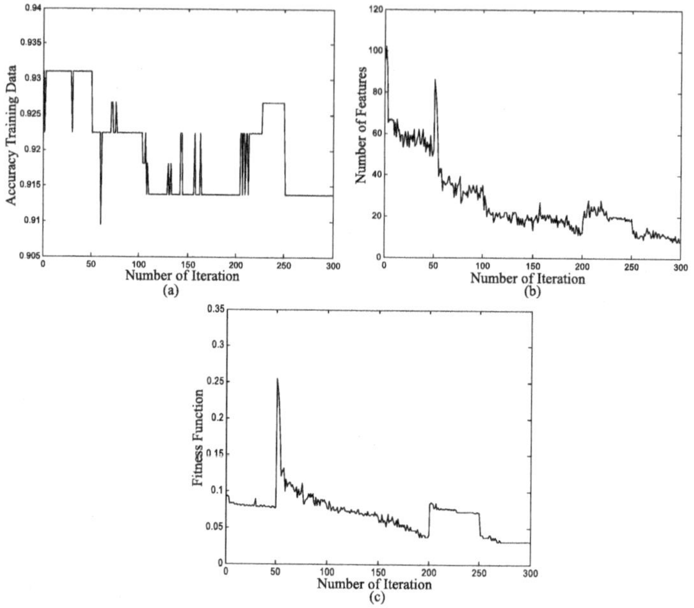

Fig. 6.12 Result of 300 iterations for GrIHS-based Ak-NN technique. (**a**) Training accuracy (**b**) Number of features (**c**) Fitness function

Table 6.7 Optimum feature subset

Feature index	Feature name	Feature nature	Feature area
26	Major axis length	Shape-based	Nucleus
64	Standard deviation	Color-based	Nucleus
119	Sum variance	Texture-based [$GLCM$ (90°)]	Whole leukocyte
161	Entropy	Texture-based [$GLCM$ (0°)]	Nucleus
270	Maximum probability	Texture-based [$GLCM$ (45°)]	Cytoplasm
273	Energy	Texture-based [$GLCM$ (90°)]	Cytoplasm

6.5.4 Comparative Study with Other Related Works Utilizing ALL Image Data

The results achieved by the proposed method have been compared with other contemporary works related to ALL detection and classification. A self-adaptive segmentation method has been used by Scotti et al. [2] and an accuracy of 92% has been achieved using ALL-IDB data set. A fuzzy-based method has been applied by Mohapatra et al. [26] and 93% accuracy has been obtained on 108 images. A combination of edge detection algorithms has been applied by Sadeghian et al. [27] on 98 ALL images and 92% accuracy was achieved. 98 ALL images have also been utilized by Madhukar et al. [28] and 93.5% accuracy was obtained. A CMYK model based leukocyte detection technique has been proposed by Putzu et al. [10] and 93.2% accuracy was reported using ALL-IDB data set. However, the entire feature set has been employed for classification purpose in the above-mentioned works [2, 10, 26–28]. They have not employed any feature selection procedure. Moreover, a bare-bone PSO-based feature reduction technique has been proposed by Srisukkham et al. [14] and 94.60% accuracy was reported. Notably, blind test accuracy has not been reported by them. They have only evaluated cross-validation accuracy. In contrast, the proposed method identifies the leukocytes by applying color-based k-means clustering on RGB color space without having any intermediate transformation phase. The proposed method achieves 95.67% leukocyte detection rate which is comparable to 93.51% detection rate proposed by Putzu et al. [10]. Moreover, a WATPSO-based effective feature selection process has been proposed which achieves 97.30% blind test accuracy utilizing only 6 effective features on the ALL-IDB2 data set. The proposed algorithm has been applied on the ALL-IDB1 data set for finding the subset of effective features and after that ALL-IDB2 data set has been used as a blind test data to prove the efficacy of the proposed algorithm. In essence, the proposed method is comparable with other related methods for leukocyte detection and ALL classification indicating the efficacy of the proposed technique. The comparative study is also shown in Table 6.8 given below.

6.6 Summary

In this chapter, a WATPSO-based feature optimization technique with the help of a dynamic input k-NN classifier is developed for enhancing the performance of ALL classification. For leukocyte identification, a color-based clustering on RGB color space has been utilized. Thereafter, nucleus and cytoplasm have been separated from the whole leukocyte. A large set of features have then been extracted from whole leukocyte, nucleus, and cytoplasm which includes shape-based, color-based, and texture-based features. Thereafter, the WATPSO-based feature optimization technique has been employed to select an effective subset of useful features which can increase the classification rate between the blast and healthy cells efficiently. The WATPSO technique utilizes two sub-swarms which can perform a heterogeneous search for two different objectives namely, increment of classification accuracy and reduction of number of features. The weights associated with different objectives transpose dynamically as the sub-swarms exchange their information between them and at the end converges to the optimized feature space. A blind testing accuracy of 97.30% on ALL-IDB2 data set has been achieved using only optimal 6 features by this methodology. The result has been compared to the other related methods which proves the efficacy of the WATPSO method. The technique can be used as a confirmatory option for hematologists to help them in the diagnostic prediction of ALL disease.

Table 6.8 Comparative study with other related works

Related works	Method	Leukocyte detection accuracy	Classification accuracy	Data set used
Scotti et al. [2]	Self-adaptive segmentation	–	92% (test accuracy)	ALL-IDB
Mohapatra et al. [26]	Fuzzy-based	–	93% (test accuracy)	108 ALL images
Sadeghian et al. [27]	Edge detection	–	92% (test accuracy)	98 ALL images
Madhukar et al. [28]	k-means clustering	–	93.5% (test accuracy)	98 ALL images
Srisukkham et al. [14]	Bare-bone PSO	–	94.6% (cross-validation accuracy)	ALL-IDB
Putzu et al. [10]	CMYK model	93.51%	93.2% (test accuracy)	ALL-IDB
–	Proposed methodology	95.67%	97.30% (blind test accuracy)	ALL-IDB

In this context, it could be summarized that proposed PSO-based techniques have performed efficiently for selecting a small subset of effective genes and effective features from gene level data set (microarray data) and cell level data set (microscopic blood smear images), respectively. Moreover, the effectiveness of this feature (gene) selection techniques have been reconfirmed by applying them to leukemia cases for both gene level and cell level data. Therefore, the proposed methodologies can be applied further to other medical image data to evaluate the efficacy of the techniques for diagnostic prediction problem. However, in the proposed techniques the solution methodology of the dual objective cost function has been executed with the help of weighted aggregation strategy. Hence, a systematic weight selection strategy can be employed to make the feature selection strategy more robust. In this context, some tissue level data such as computed tomography images are considered to evaluate the diagnostic prediction of cancer in the next chapter.

References

1. Sinha N, Ramakrishnan A (2003) Automation of differential blood count. In: TENCON 2003. Conference on Convergent Technologies for Asia-Pacific Region, vol. 2. IEEE pp 547–551
2. Scotti F (2006) Robust segmentation and measurements techniques of white cells in blood microscope images. In: 2006 IEEE Instrumentation and Measurement Technology Conference Proceedings. IEEE, pp 43–48
3. Mohapatra S, Patra D, Satpathy S (2014) An ensemble classifier system for early diagnosis of acute lymphoblastic leukemia in blood microscopic images. Neural Comput Appl 24:1887
4. Chin Neoh S, Srisukkham W, Zhang L, Todryk S, Greystoke B, Peng Lim C, Alamgir Hossain M, Aslam N (2015) An intelligent decision support system for leukaemia diagnosis using microscopic blood images. Sci Rep 5:14938
5. Piuri V, Scotti F (2004) Morphological classification of blood leucocytes by microscope images. In: 2004 IEEE International Conference on Computational Intelligence for Measurement Systems and Applications. CIMSA. IEEE, pp 103–108
6. Madhloom HT, Kareem SA, Ariffin H (2012) A robust feature extraction and selection method for the recognition of lymphocytes versus acute lymphoblastic leukemia. In: 2012 International Conference on Advanced Computer Science Applications and Technologies (ACSAT). IEEE, pp 330–335
7. Rezatofighi SH, Soltanian-Zadeh H (2011) Automatic recognition of five types of white blood cells in peripheral blood. Comput Med Imaging Graph 35:333
8. Agaian S, Madhukar M, Chronopoulos AT (2014) Automated screening system for acute myelogenous leukemia detection in blood microscopic images. IEEE Syst J 8:995
9. Singhal V, Singh P (2014) Local binary pattern for automatic detection of acute lymphoblastic leukemia. In: 2014 Twentieth National Conference on Communications (NCC). IEEE, pp 1–5
10. Putzu L, Caocci G, Di Ruberto C (2014) Leucocyte classification for leukaemia detection using image processing techniques. Artif Intell Med 62:179
11. Huang D-C, Hung K-D (2012) Leukocyte nucleus segmentation and recognition in color blood-smear images. In: 2012 IEEE International Instrumentation and Measurement Technology Conference Proceedings. IEEE, pp 171–176
12. Osowski S, Siroic R, Markiewicz T, Siwek K (2008) Application of support vector machine and genetic algorithm for improved blood cell recognition. IEEE Trans Instrum Meas 58:2159

13. Escalante HJ, Montes-y Gómez M, González JA, Gómez-Gil P, Altamirano L, Reyes CA, Reta C, Rosales A (2012) Acute leukemia classification by ensemble particle swarm model selection. Artif Intell Med 55:163
14. Srisukkham W, Zhang L, Neoh SC, Todryk S, Lim CP (2017) Intelligent leukaemia diagnosis with bare-bones PSO based feature optimization. Appl Soft Comput 56:405
15. Gonzalez R, Gonzalez RC, Woods RE (2002) Digital image processing, 2nd edn. Prentice Hall, Upper Saddle River, NJ
16. Kar S, Das Sharma K, Maitra M (2022) Optimised feature selection for identification of carcinogenic leukocytes using weighted aggregation based transposition PSO. IETE J Res 68:1991
17. Kar S, Sharma KD, Maitra M (2015) Gene selection from microarray gene expression data for classification of cancer subgroups employing PSO and adaptive K-nearest neighborhood technique. Expert Syst Appl 42:612
18. Haralick RM, Shanmugam K, Dinstein IH (1973) Textural features for image classification. IEEE Trans Syst Man Cybern 6:610–621
19. Eberhart R, Kennedy J (1995) A new optimizer using particle swarm theory. MHS'95. Proceedings of the sixth international symposium on micro machine and human science. A new optimizer using particle swarm theory. IEEE, pp 39–43
20. Wang S, Phillips P, Yang J, Sun P, Zhang Y (2016) Magnetic resonance brain classification by a novel binary particle swarm optimization with mutation and time-varying acceleration coefficients. Biomed Eng/Biomedizinische Technik 61:431
21. Zhang Y, Wang S, Sui Y, Yang M, Liu B, Cheng H, Sun J, Jia W, Phillips P, Gorriz JM (2018) Multivariate approach for Alzheimer's disease detection using stationary wavelet entropy and predator-prey particle swarm optimization. J Alzheimer's Dis 65:855
22. Ishibuchi H, Setoguchi Y, Masuda H, Nojima Y (2016) Performance of decomposition-based many-objective algorithms strongly depends on Pareto front shapes. IEEE Trans Evol Comput 21:169
23. Parsopoulos KE, Vrahatis MN (2008) Multi-objective optimization in computational intelligence: theory and practice. IGI Global, Hershey, pp 20–42
24. Labati RD, Piuri V, Scotti F (2011) All-IDB: The acute lymphoblastic leukemia image database for image processing. 2011 18th IEEE international conference on image processing. IEEE, pp 2045–2048
25. Vania M, Mureja D, Lee D (2017) Automatic spine segmentation using convolutional neural network via redundant generation of class labels for 3D spine modeling. Preprint. arXiv:1712.01640
26. Mohapatra S, Samanta SS, Patra D, Satpathi S (2011) Fuzzy based blood image segmentation for automated leukemia detection. 2011 international conference on devices and communications (ICDeCom). IEEE, pp 1–5
27. Sadeghian F, Seman Z, Ramli AR, Abdul Kahar BH, Saripan M-I (2009) A framework for white blood cell segmentation in microscopic blood images using digital image processing. Biol Proced Online 11:196
28. Madhukar M, Agaian S, Chronopoulos AT (2012) New decision support tool for acute lymphoblastic leukemia classification. Image processing: algorithms and systems X; and parallel processing for imaging applications II, vol 8295. SPIE, pp 367–378

Chapter 7
Intelligent Computing Approach for Lung Nodule Detection

Abstract This chapter focuses on automating the diagnosis of lung cancer by identifying and classifying lung nodules in lung CT images. This approach involves formulating a weighted dual goal optimization problem that selects an optimal feature subset for accurate classification of lung nodules. In order to guarantee that a compact feature subset enhances nodule classification accuracy, the weights allocated to the discriminating feature subset and classification accuracy are strategically determined. Here, an adaptive weighted aggregation strategy based on the Group Improvised Harmony Search (GrIHS) evolutionary optimization technique has been proposed to tackle the optimization problem. To further improve nodule classification accuracy, an adaptive k-NN classifier has been integrated into the GrIHS technique to determine the optimal number of neighborhoods in the search space. The proposed methodology has been successfully applied to the Lung Image Database Consortium (LIDC) data set. The experimental findings demonstrate the efficacy of the proposed method, with only 12 differentiating features required to achieve a remarkable sensitivity of 97.59% and blind testing accuracy of 97.78%. Consequently, the proposed methodology has the potential to assist radiologists in diagnosing lung cancer by providing reliable support in the interpretation of lung computed tomography images.

7.1 Introduction

In the previous chapters, the methodologies have been developed for the detection and classification of cancers using gene level data and cell level data. In this chapter, the detection and classification of cancers from tissue level data has been given the main focus. In this context, computed tomography (CT) plays the vital role. CT is a sensitive image modality for the tissue level images. In this work, the detection and classification of lung cancer using the lung CT image has been discussed because the mortality rate of lung cancer is quite high [1, 2]. Lung cancers associated with various form of prognostic natures, are still believed to provide a good number of event free survival provided the cancerous nodules

K. Das Sharma et al., *Intelligent Computing in Carcinogenic Disease Detection*, Computational Intelligence Methods and Applications,
https://doi.org/10.1007/978-981-97-2424-6_7

could be detected in an early stage. Manual interventions for diagnosing such a probabilistically terminal disease are being practiced. However, a paradigm shift as reported in the literature has proved that happening image processing tools and methodologies can supplement the empirical experience of the pathologists with the help of artificially computational procedure with almost near-perfect accuracy.

In the literature, different lung segmentation and nodule detection techniques have been reported [3–22]. Several approaches have also been reported to enhance the performance of lung nodule classification by analyzing different categories of features [23–27]. Recently, stochastic optimization-based feature selection methodologies for different applications have been reported in the literature [28–35].

Therefore, the motivation of the proposed work lies in the fact that the performance of lung nodule classification can further be improved by selecting a small discriminant feature subset through investigating a large number of features from the detected nodule-like structures. In this work, a group improvised harmony search (GrIHS) methodology has been utilized to identify a small subset of discriminating features. An adaptive k-NN classifier has also been embedded with GrIHS technique to select a proper value of neighborhood that would help to enhance the classification accuracy. The adaptive weighted aggregation strategy has been proposed to select the weights adaptively that are associated with discriminating features and classification accuracy [36]. The proposed weight selection strategy would provide an optimal combination of discriminating features and classification accuracy in the search space to guide the algorithm toward enhancing the classification performance. In this work, the blind testing samples have been completely excluded from the GrIHS-based feature selection technique to alleviate the selection bias. However, blind testing has been done after the GrIHS-based feature selection process is complete to evaluate the robustness of our proposed scheme. Therefore, the contributions of this work are summarized below:

- Utilization of the GrIHS tool for identifying the most informative subset of features.
- Adoption of A k-NN classifier interweaved with the GrIHS tool (for evaluating the optimized value of k to be used for enhancing the classification accuracy).
- Last but not the least, proposition of a novel adaptive weight selection strategy for obtaining a solution for the dual objective problem as stated. The challenge, as said earlier, is to decrease the number of informative features vis-a-vis increasing the classification accuracy.

Hence, we emphasize that this work presents an adaptive weight selection approach intertwined with the GrIHS-based feature selection method toward enhancing the classification performance in lung nodule detection that can be used as a reliable decision support system for radiologists for accurate detection of lung nodules.

7.2 Preprocessing

In this work, first, the lung region has been separated from other body parts using the Hounsfield Unit (HU) and mathematical morphology-based technique [37]. Then the contour correction method [38] has been employed to include juxta-pleural nodules (i.e., nodules that touched with lung wall) in the lung mask. The surface normal-based volumetric shape index [39] has then been calculated for each voxel within the segmented lungs to enhance the objects of specific shapes. For accurate calculation of the shape index-based geometric features, Gaussian smoothing have also been performed prior to the shape index calculation. Thereafter, the multi-scale dot enhancement filtering [40] has been performed to identify the potential lung nodule candidates (PLNCs).

7.2.1 Lung Segmentation

The objective of lung segmentation is to separate the entire lung region from other anatomical portion of the body. Generally, the lung region is a bag of air inside the body. Therefore, the lung shows up as a dark region in the CT scan image. This disparity between lung region and other tissues are the basis of the lung segmentation scheme. In this work, an HU-based simple technique has been applied. Here, the lung region has been extracted from the image by using a threshold of −400HU, the Hounsfield unit of air [37]. The initial lung masked image has been obtained by performing border removing operation on the thresholded image. Then the complete lung masked image has been obtained by performing hole filling operation on the initial lung masked image. The complete lung masked image, thus obtained, does not include the nodules that touch the lung border, i.e., juxta-pleural nodules. These juxta-pleural nodules can affect nodule detection performance. Therefore, a contour correction technique [38] has been employed in this work to include the juxta-pleural nodule in the segmented lung region for enhancing the performance of the nodule detection methodology. The HU-based lung segmentation methodology has been shown in Algorithm 7.1. Each step of the lung segmentation technique has been shown in Fig. 7.1.

7.2.2 Volumetric Shape Index

The volumetric shape index represents the topological shape of the volume in the neighborhood of a voxel. It maps the points in a surface to the linear scale $[-1, 1]$. In this work, the shape index of each point in the lung surface has been calculated by considering the local differential structure of surfaces in terms of surface normal. Let $p\,(x, y, z)$ denote a point on the lung surface. At each point p, there exists a

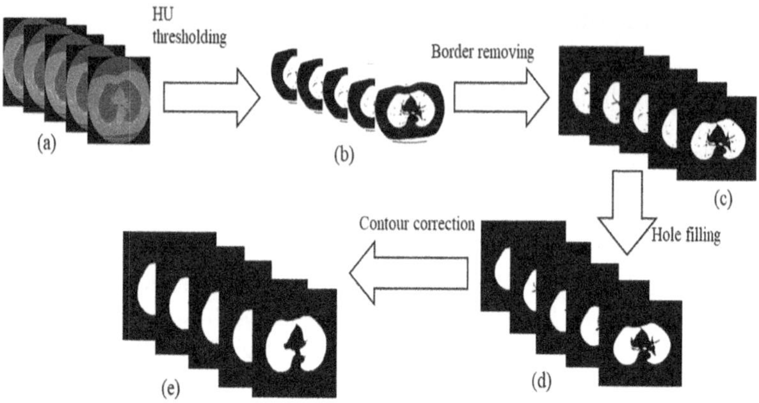

Fig. 7.1 HU-based lung segmentation. (**a**) Original lung CT image (**b**) HU-based threshold image (**c**) Initial lung masked image (**d**) complete lung masked image (**e**) Contour correction

---

Algorithm 7.1 Lung Segmentation

---

for the entire image **do**
  if $I_{CT} \leq -400\text{HU}$ **then**
    obtain HU-based threshold image
  end if
  obtain initial lung mask
  obtain complete lung mask
  obtain contour correction
end for

---

small neighborhood of p, in which z can be expressed by a function, $z = f(x, y)$. Then the differential structure of the surface has been captured by the local Hessian matrix as [39]:

$$\mathcal{H} = \begin{bmatrix} \frac{\partial^2 z}{\partial x^2} & \frac{\partial^2 z}{\partial x \partial y} \\ \frac{\partial^2 z}{\partial y \partial x} & \frac{\partial^2 z}{\partial y^2} \end{bmatrix} \tag{7.1}$$

The two eigenvalues of the $\mathcal{H}$ are the maximum (c_{max}) and minimum (c_{min}) curvatures. The mean curvature (C) of the surface is calculated by averaging the value of maximum (c_{max}) and minimum (c_{min}) curvatures.

$$C = \frac{1}{2}(c_{max} + c_{min}) \tag{7.2}$$

The Gaussian curvature ($\mathcal{G}$) of the surface can be calculated by the product of the two eigenvalues.

$$\mathcal{G} = c_{max} c_{min} \tag{7.3}$$

To define the lung surface in terms of surface normal, the set of surface normals, $n = \left(\frac{\partial z}{\partial x}, \frac{\partial z}{\partial y}, 1\right)^T$, have been substituted in Eq. (7.1) to represent the local Hessian matrix as:

$$\mathcal{H} = \begin{bmatrix} \left(\frac{\partial n}{\partial x}\right)_x & \left(\frac{\partial n}{\partial x}\right)_y \\ \left(\frac{\partial n}{\partial y}\right)_x & \left(\frac{\partial n}{\partial y}\right)_y \end{bmatrix} \tag{7.4}$$

where $(.)_x$ and $(.)_y$ denote the x and y components of the parenthesized vectors, respectively. As the mean curvature and Gaussian curvature of the lung surface are the eigenvalues of $\mathcal{H}$, the maximum and minimum curvatures have been expressed as:

$$c_{max} = -\frac{1}{2}\left[\left(\frac{\partial n}{\partial x}\right)_x + \left(\frac{\partial n}{\partial y}\right)_y - \sqrt{\left(\left(\frac{\partial n}{\partial x}\right)_x - \left(\frac{\partial n}{\partial y}\right)_y\right)^2 + 4\left(\frac{\partial n}{\partial x}\right)_y \left(\frac{\partial n}{\partial y}\right)_x}\right] \tag{7.5}$$

$$c_{min} = -\frac{1}{2}\left[\left(\frac{\partial n}{\partial x}\right)_x + \left(\frac{\partial n}{\partial y}\right)_y + \sqrt{\left(\left(\frac{\partial n}{\partial x}\right)_x - \left(\frac{\partial n}{\partial y}\right)_y\right)^2 + 4\left(\frac{\partial n}{\partial x}\right)_y \left(\frac{\partial n}{\partial y}\right)_x}\right] \tag{7.6}$$

Thus the volumetric shape index (SI) at point p has been expressed in terms of surface normal [41] as:

$$\begin{aligned} SI(p) &= \frac{2}{\pi} \arctan \frac{c_{max} + c_{min}}{c_{max} - c_{min}} \\ &= \frac{2}{\pi} \arctan \left[\frac{\left(\frac{\partial n}{\partial x}\right)_x + \left(\frac{\partial n}{\partial y}\right)_y}{\sqrt{\left\{\left(\frac{\partial n}{\partial x}\right)_x - \left(\frac{\partial n}{\partial y}\right)_y\right\}^2 + 4\left(\frac{\partial n}{\partial x}\right)_y \left(\frac{\partial n}{\partial y}\right)_x}} \right] \end{aligned} \tag{7.7}$$

The volumetric SI values can be mapped on a unit circle. The unit circle containing different SI values, i.e., different shapes of each point on lung surface is shown in Fig. 7.2. The "negative" values of SI represent shapes that are each other's

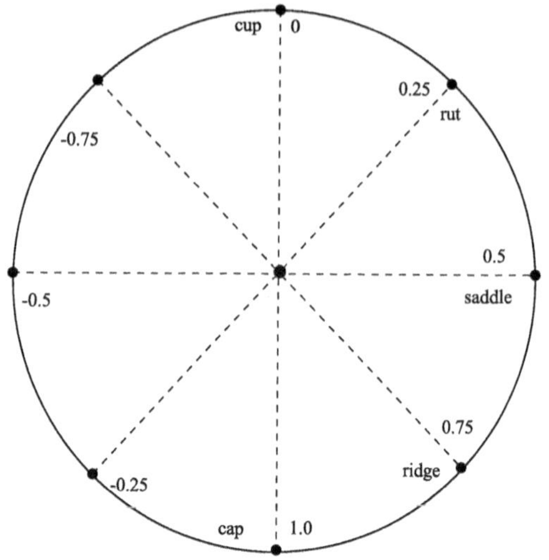

Fig. 7.2 Relationship of different shapes to the SI values

"negative," i.e., they have the same shape, but the opposite growth [42]. Therefore, all shapes of $SI \in [-1, 1]$ can be remapped on the interval $SI \in [0, 1]$. For example, five well-known shape classes have the following shape index values: cup (0.0), rut (0.25), saddle (0.5), ridge (0.75), and cap (1.0). The SI value of 1.0 represents a "cap" which indicates a sphere-like shape (nodule) [38]. The SI value of 0.75 represents a "ridge" which indicates a cylinder-like shape (blood vessel) [38]. Therefore, it can be noticed that the SI values for sphere-like nodules are generally higher than the values for cylinder-like blood vessels. This dissimilarity in SI values can be used as a discriminating feature between lung nodules (sphere-like) and blood vessels (cylinder-like).

In this chapter, a lung nodule candidate (LNC) extraction algorithm using the SI values has been applied. The points in a lung region are detected as LNCs if they satisfy Eq. (7.8). The extracted LNCs have been shown in Fig. 7.3 where the original lung nodules are shown in white circle. It is observed that some LNCs are identified inside the white circle. This demonstrates that each lung nodule contains, at least, one point that can be sensed by the SI map.

$$SI \geq \psi_1, SI \leq \psi_2 \tag{7.8}$$

where ψ_1 and ψ_2 are high and low SI threshold chosen experimentally.

7.2.3 *Multi-scale Dot Enhancement Filter*

Lung nodules are often detected using SI values with a large number of false positives(FPs). To reduce the number of FPs, the selective enhancement filter is used

Algorithm 7.2 LNC extraction

for each point, $p(x, y, z)$ **do**
  if $SI \geq \psi_1$ and $SI \leq \psi_2$ **then**
    $I(p)$=LNC
  end if
end for

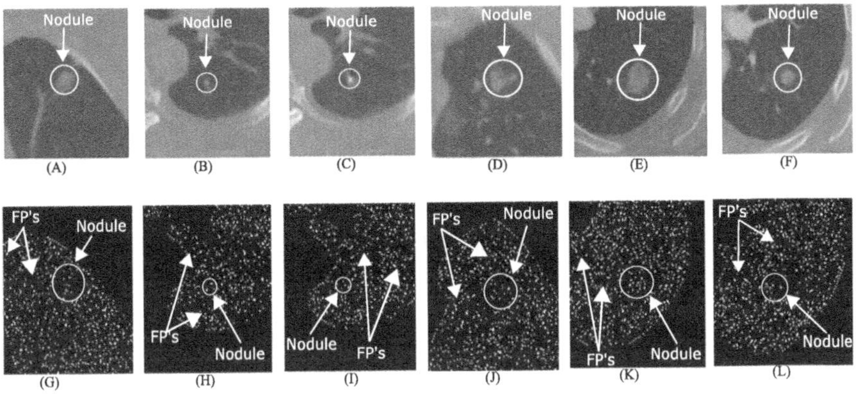

Fig. 7.3 Extraction of LNCs

to enhance the objects of a specific shape. In this work, the aim is to enhance the dot-like structures (such as lung nodules) and suppress the other alike structures (such as blood vessels) from the lung structure. Therefore, a multi-scale dot enhancement filter has been used in the shape index-based image to enhance the LNCs and suppress the FPs effectively. A dot and the Hessian matrix in the lung CT image can be expressed as [40]:

$$dot(x, y, z) = exp\left\{-\frac{x^2 + y^2 + z^2}{2\sigma^2}\right\} \tag{7.9}$$

$$\mathcal{H} = \begin{bmatrix} I_{xx} & I_{xy} & I_{xz} \\ I_{yx} & I_{yy} & I_{yz} \\ I_{zx} & I_{zy} & I_{zz} \end{bmatrix} \tag{7.10}$$

where σ is the scale of the dot and $I_{xx}, I_{xy}, I_{xz}, I_{yx}, I_{yy}, I_{yz}, I_{zx}, I_{zy}, I_{zz}$ are the second derivatives of the lung image. The eigen values of $\mathcal{H}$, λ_1, λ_2, and λ_3 must satisfy the following equations for a dot and a line, respectively [40]:

$$\lambda_1 = \lambda_2 = \lambda_3 = \frac{1}{\sigma^2} < 0 \tag{7.11}$$

$$\lambda_1 = \lambda_2 = \frac{1}{\sigma^2} < 0, \lambda_3 = 0 \tag{7.12}$$

To enhance the dot-like structures, a dot must be distinguished clearly from line. Therefore, a unique distinguishing factor is needed which would be satisfied by the dot only. Now, let

$$e_2 = \frac{|\lambda_2|}{|\lambda_1|}, e_3 = \frac{|\lambda_3|}{|\lambda_1|} \tag{7.13}$$

So, the above condition in Eq. (7.13) for dot becomes

$$e_2 = 1, e_3 = 1 \tag{7.14}$$

and the condition in Eq. (7.13) for line becomes

$$e_2 = 1, e_3 = 0 \tag{7.15}$$

Now as $e_2 = 1$ for both dot and line, it can be rejected as a distinguishing factor. Therefore, e_3 become the only unique distinguishing factor between a dot and a line. So the likelihood of a dot has been defined as:

$$\kappa_{dot}(\lambda_1, \lambda_2, \lambda_3) = e_3 = \frac{|\lambda_3|}{|\lambda_1|} \tag{7.16}$$

The value of $|\lambda_3|$ provides a value greater than 0 for a dot. Therefore, the magnitude can be defined as:

$$g_{dot}(\lambda_1, \lambda_2, \lambda_3) = |\lambda_3| \tag{7.17}$$

Therefore, the output $Z_{dot}(\lambda_1, \lambda_2, \lambda_3)$ for a dot will be [40]:

$$\begin{aligned} Z_{dot}(\lambda_1, \lambda_2, \lambda_3) &= g_{dot}(\lambda_1, \lambda_2, \lambda_3)\, \kappa_{dot}(\lambda_1, \lambda_2, \lambda_3) \\ &= \frac{|\lambda_3|^2}{|\lambda_1|}, \text{if } \lambda_1 < 0, \lambda_2 < 0, \lambda_3 < 0 \\ &= 0, \text{otherwise} \end{aligned} \tag{7.18}$$

However, the multi-scale dot enhancement filter has been used here to reduce the effect of image noise and to take into account different object sizes by performing Gaussian image smoothing considering different scales. The multi-scaling has been performed before the calculation of second derivatives by Gaussian smoothing with different scales.

In this work, the diameter of the nodule object to be enhanced is in a range of $[d_0, d_1] \approx [3\,\text{mm}, 30\,\text{mm}]$. The N discrete smoothing scales σ_N in the range of $\left[\frac{d_0}{4}, \frac{d_1}{4}\right]$ can be calculated as:

$$\sigma_1 = \frac{d_0}{4}, \sigma_2 = r\sigma_1, \ldots, \sigma_N = r^{N-1}\sigma_1 \tag{7.19}$$

Algorithm 7.3 PLNC extraction

calculate individual smoothing scales using Eq. (7.19)
for each smoothing scales $\sigma_1, \sigma_2, \ldots, \sigma_N$ **do**
  for each point $p\,(x, y, z)$ **do**
    compute the Hessian matrix using Eq. (7.10)
    compute its eigenvalues λ_1, λ_2 and λ_3
    compute $Z_{dot}\,(\lambda_1, \lambda_2, \lambda_3)$ using Eq. (7.18)
  end for
end for
compute the final dot value by selecting the maximum dot value calculated from the different smoothing scales

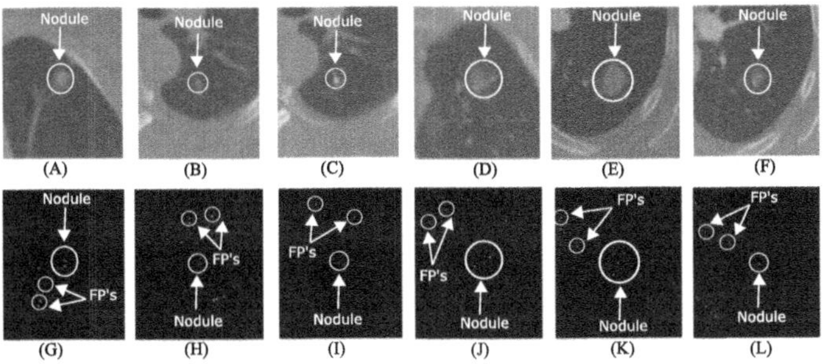

Fig. 7.4 Multi-scale dot enhancement filtering

where $r = \left(\frac{d_1}{d_0}\right)^{\frac{1}{N-1}}$. The maximum dot value calculated using Eq. (7.18) among the different scales in Eq. (7.19) has been chosen to be the final dot value for each voxel. From the SI map, a dot is kept as a PLNC, if the dot value is greater than a predefined threshold value chosen experimentally. In this work, a PLNC extraction algorithm based on the multi-scale dot enhancement filter has been applied to identify the PLNCs. Figure 7.4 shows the examples of PLNC extraction based on shape index map and multi-scale dot enhancement filtering.

7.3 Lung Nodule Detection and Classification Methodology

In this work and as per our proposition, first, sub-images have been extracted from each PLNC. Then, different types of features such as shape-based features, color-based features, and texture-based features have been extracted from each sub-image to construct a feature pool. The shape-based features have been derived from shape index map shown in Table 7.1. The color-based features have been derived from the intensity distribution of the sub-image regions. These features have been shown in Table 6.2. However, in this work, not only the intensity distribution of the sub-

Table 7.1 Shape-based features

Index no.	Feature name	Description
1	Max of shape index map	Maximum shape index value of the region
2	Min of shape index map	Minimum shape index value of the region
3	Mean of shape index map	Mean shape index value of the region
4	Standard deviation of shape index map	Standard deviation of the shape index value of the region

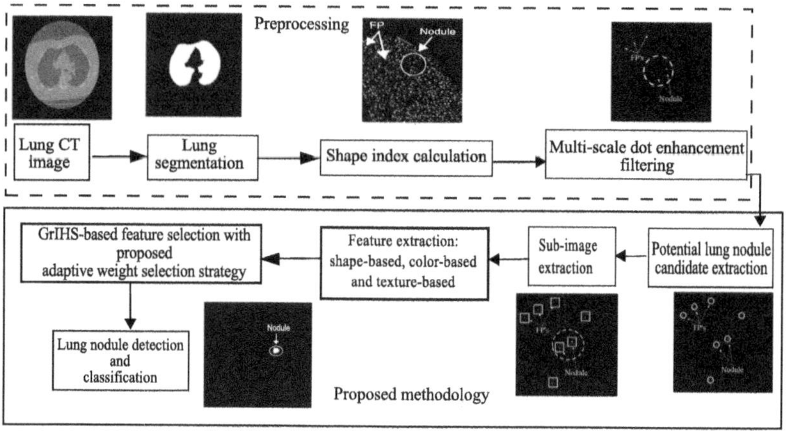

Fig. 7.5 Block diagram of the methodology

image regions have been considered but also the positional information of the pixels having identical gray value have also been considered. These information are usually carrying the texture features and have been calculated from GLCM. The adjacency of the pixels in Eq. (6.3) can be judged for angles 0°, 45°, 90° and 135° as shown in Fig. 6.7. Consequently, GLCM features enumerated in Table 6.3 are evaluated for the four angles resulting in a total of 80 GLCM feature.

Therefore, a total number of 90 features (shape: 4, color: 6 and texture: 80, as shown in Tables 7.1, 6.2 and 6.3) have been considered in this work. This large number of features sustain severe computational burden. However, the large set of features contain some redundant features which may reduce the classification accuracy of the lung nodules. Therefore, identification of an important subset of useful features is necessary to achieve better classification accuracy. In this work, a GrIHS-based feature selection technique has been used to achieve the above-mentioned task. The weights that are associated with selected features and classification accuracy has also been proposed to be selected adaptively to further improve the classification accuracy. The block diagram of the methodology is shown in Fig. 7.5 where the preprocessing block is shown in dotted line and the proposed block is shown in solid line.

7.3.1 Harmony Search Algorithm

Harmony search algorithm, a relatively recently proposed meta-heuristic optimization algorithm, first proposed by Z. W. Geem in 2001 [43]. It was conceptualized using the musical process of searching for a perfect state of harmony, based on western classical music. It is one of the few optimization algorithms not inspired by any natural phenomenon. The harmony in music can be modeled as analogous to a candidate solution vector in search space. The musician improvisation bears strong semblance with the search techniques in an optimization strategy. One of the many features of the HS algorithm is that it does not require precise initialization of the decision variables [43]. Moreover, in comparison to a gradient search method, the HS algorithm utilizes a stochastic search that is based on the three music inspired operators, namely, harmony memory considering rate ($HMCR \in [0, 1]$, rate of the value of the variable selection from the musician's memory), pitch adjustment rate ($PAR \in [0, 1]$), rate of the fine-tuning of the value of the variable selected from memory) and band width (bw, the distance between upper and lower bound for each variable). However, HS has a stochastic derivative [44] which uses the musician's experience-based derivative information instead of calculus-based one. In comparison to the other stochastic optimization techniques, the HS requires fewer mathematical steps (some IF-THEN rules are to be calculated) and easily implementable. As the HS algorithm does not require any information about the process to be optimized it can be easily applied to various types of engineering optimization problems.

In the HS algorithm, a harmony memory (*HM*) is first initialized uniformly at random. Then utilizing *HMCR* and *PAR* a new harmony will be generated in each harmony improvisation stage (optimization iteration). The aesthetic quality of the newly generated harmony is evaluated using a fitness function and the harmony memory will again be sorted based on that fitness value rejecting the worst one. The basic HS algorithm has been improvised by manipulating the controlling parameter of the algorithm, i.e., *HMCR*, *PAR* etc. [43]. The basic HS algorithm is shown in Algorithm 7.4.

In this work, the problem has been defined as to find a subset of useful features that would improve the nodule classification accuracy. The solution methodology rests on GrIHS-based feature selection algorithm. The basic HS algorithm utilizes fixed value of *HMCR*, *PAR*, and *bw* to guide the algorithm in the solution space. M. Mahdavi et al. [45] have proposed a modification over the basic HS algorithm by varying *PAR* and *bw* throughout the optimization and keeping *HMCR* constant. Another variant of the HS algorithm has been proposed by K. Das Sharma et al. [46], where *HMCR* and *PAR* parameters are varied simultaneously throughout the optimizing period keeping *bw* value constant. K. Das Sharma has also proposed another variant of HS in [47] where an *lbest* topological model was introduced inspired by the philharmonic orchestration in western classical music composition. The physical implication of incremental variation of *HMCR* parameter throughout the harmony improvisation stage is that as the generation of harmony improvisation increases

Algorithm 7.4 Basic HS algorithm

Define objective function $g(z)$, $z = [z_1 z_2 \ldots z_n]$; n=number of decision variable
Define *HMS*, *HMCR*, *PAR* and *bw*
Generate *HM* with random *HMS* number of harmonies
while $t < maxnumberofiterations$ **do**
  while $i \leq numberofvariablesn$ **do**
    if $rand < HMCR$ **then**
      Choose a value from HM for the variable i
      if $rand < PAR$ **then**
        $z_i = z_i + bw \times ud(-1, +1)$, ud is uniformly distributed function
      end if
    else
      Choose a random value
    end if
  end while
  Accept the new harmony (candidate solution vector) if better.
end while
Select the current best solution as optimum solution

the HS algorithm relies more on the harmony memory since the harmony memory would be rich in experience owing to the improvisations during the generations. The *PAR* parameter is also augmented because the fine-tuning is necessary as the HS algorithm relies more on the previous values of the harmonies accumulated in the HM [46]. Another parameter group memory considering rate (*GMCR*) is also introduced with the concept of group-wise improvisation of harmony memory. In this improved version of the HS algorithm, the harmony memory is divided into some groups and the member harmonies for each such group are randomly chosen from HM in each iteration [47]. The logic behind this modification is that, a particular harmony may exist in different groups in different iterations and that makes the exploration of the solution space in a better way and that in effect tend the solution toward global optima. The difference between the basic HS algorithm and GrIHS algorithm during the improvisation stage is that in earlier case only one harmony improvisation will occur in each iteration whereas in GrIHS model the number of harmony improvisation will depend on the number of groups selected in each iteration.

7.3.2 *Fitness Function*

The objectives of this work are two-folds. The aim is to increase the classification accuracy by reducing the subset size that contains the effective non-redundant features. To tackle this dual objective, the weighted aggregation approach has been taken into consideration toward the solution. In this approach, the solution has been obtained by systematically assigning the weights to each objective. The fitness function in the weighted aggregation approach has been constructed in such a way so as to achieve the dual objectives. Therefore, the fitness function has been chosen as:

$$FF = E_{CV} \times \omega_1 + \left(\frac{n_{fsubset}}{T_f}\right) \times \omega_2 \tag{7.20}$$

where E_{CV} is the error over cross-validation accuracy. Here, E_{CV} has been calculated as:

$$E_{CV} = 1 - CV_{Accuracy} \tag{7.21}$$

where $CV_{Accuracy} \in [0, 1]$. $n_{fsubset}$ is the number of effective features in the candidate feature subset. T_f is the total number of features. ω_1 and ω_2 are the two weights corresponding to the different objectives. The performance of the fitness function depends on the proper selection of the weights. Thus, an adaptive weight selection strategy has been proposed for the optimal combination of classification accuracy and selected features.

7.3.3 Adaptive Weight Selection Strategy

In this work, the two objectives have been combined together by weight factors. Several solutions can be obtained by slightly changing the set of weight factors. Generally, traditional weight selection strategy follows the researcher's experience. Thus, a methodical and logical adaptive weight selection strategy has been proposed that demonstrates the optimal solution point that can be chosen from a set of weights. The weights have been assigned to define the relative importance of the objectives. Thus, the suggested weight selection strategy treats each objective with similar magnitudes and does not allow any objective to dominate the aggregated function. An attempt has been made to select the weights based on the standard deviation and mean value of the objectives due to the random variation of the input variables. The lower standard deviation of objectives indicates the weak bonding with the variable. Therefore, higher weight has been assigned with it to get improved compromising solution. The weights of different objectives have been calculated as follows:

$$\omega_i = |s_i + r \times (m_i - s_i)| \tag{7.22}$$

where

$$s_i = \frac{1/\psi_i}{\sum_{i=1}^{N_{obj}} 1/\psi_i} \tag{7.23}$$

$$m_i = \frac{1/\mu_i}{\sum_{i=1}^{N_{obj}} 1/\mu_i} \tag{7.24}$$

r is the random number between 0 and 1. ψ_i and μ_i represent the standard deviation and mean value of the ith objective, respectively, and N_{obj} represents the total number of objectives present in the fitness function. In the proposed adaptive weight selection strategy, the objectives are adaptively weighted and aggregated to converge at a well optimal solution.

7.3.4 GrIHS-Based Feature Selection Strategy

In this work, the feature selection strategy has been implemented using the concept of group improvisation of the HS algorithm. The k-NN classifier has also been allowed to embed with the GrIHS technique. As finding a proper value of neighborhood is a real challenge for k-NN-based classification, the proposed adaptability of k-NN using group improvised HS helps to tackle the problem in a logical way. Here, GrIHS has been allowed to search over the entire feature space defined by the PLNCs. A common threshold value has been chosen to be associated with each harmony to characterize the candidate solution. The choice between inclusion and exclusion of a feature in the subset completely depends on each harmony that has been binarized by that suitable threshold value.

In this work, the HM is constructed using harmony memory size (HMS) as:

$$HM = \left[\begin{array}{cccc|c} x_1^1 & x_2^1 & \ldots & x_n^1 & f\left(x^1\right) \\ x_1^2 & x_2^2 & \ldots & x_n^2 & f\left(x^2\right) \\ \ldots & \ldots & \ldots & \ldots & \\ x_1^{HMS} & x_2^{HMS} & \ldots & x_n^{HMS} & f\left(x^{HMS}\right) \end{array}\right] \tag{7.25}$$

where f (.) is the fitness function. The initial HM has been populated by real numbers (0, 1). The threshold ϵ has been employed to construct the gene selection flag of binary bits according to Eq. (6.9). Each element of a feature selection flag bears the information of existence or nonexistence of a feature in its corresponding candidate solution vector. In a binary harmony, '0' indicates the nonexistence and '1' corresponds to the existence of the corresponding feature from a feature pool. The total number of '1's and its position in the feature selection flag keeps changing in each harmony improvisation and forms the effective feature subset of the entire feature pool.

To differentiate lung nodules from the healthy part of the lung in an efficient manner some useful features have to be selected and that features must have the adequate discriminatory power to distinguish the nodules from the non-nodules [48]. To reach the above-mentioned goal, first, the CV training accuracy and CV test accuracy have been calculated using Eqs. (5.1) and (5.2). For each k, k number of single estimations

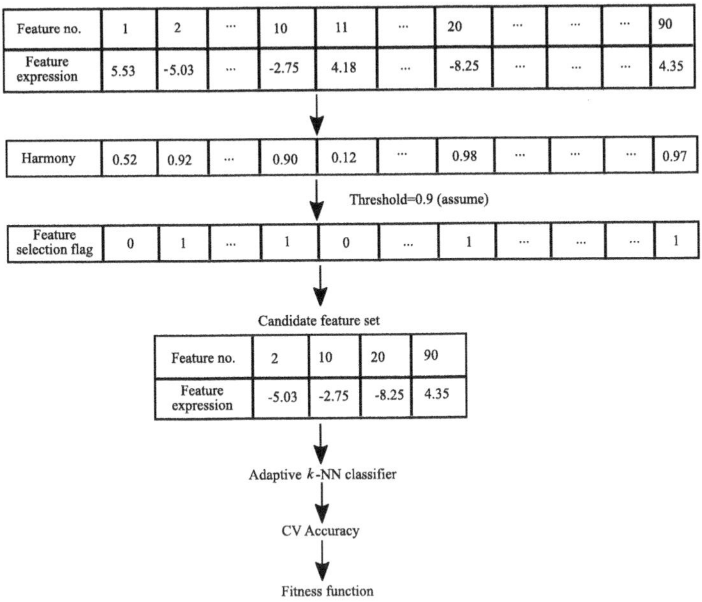

Fig. 7.6 Simplified diagram of proposed optimized feature-based classification process of lung nodules

(SE) have then been calculated using Eq. (5.3). From the k number of single estimations, the CV accuracy has been calculated as: $max\ (SE\ (K))$. Therefore, the fitness function for the proposed FP reduction technique can be constructed using Eq. (7.20). In this proposed approach, ω_1 and ω_2 have been selected by the proposed adaptive weight selection strategy. This weight selection strategy would help to further improve the classification accuracy with small number of discriminating features. As the testing result could be overoptimistic if the blind testing samples are not excluded from the training process, the proposed GrIHS-based feature selection algorithm totally excludes the blind testing samples in order to avoid the influence of selection bias. The simplified diagram of our method is shown in Fig. 7.6 and the algorithm of the proposed technique is shown in Algorithm 7.5.

7.4 Experimental Study

In this work, 227 number of CT scans from the LIDC database [49] have been used to assess the efficacy of the proposed methodology. Each scan of the database was read individually by four radiologists of a qualified panel. Then a consensual ground truth was defined by the panel [49].

The 227 scans have been processed by the proposed methodology. LNCs have been first extracted from the SI map, where SI threshold has been set to be $0.9 \leq$

Algorithm 7.5 GrIHS-based feature selection algorithm

BEGIN *%start the algorithm*
Define objective function $g(z)$, $z = [z_1\ z_2\ \ldots\ z_n]$; n=number of decision variable *%define the objective function*
Define HMS, $HMCR_{max}$, $HMCR_{min}$, PAR_{max}, PAR_{min}, bw, $GMCR$, G_{max} and G_{min} *% define the parameters*
Generate HM with random HMS number of harmonies *% define the harmony memory*
while $t < max\ number\ of\ iterations$ **do**
  //create a new solution in the following way//
  //feature selection flag//
  for each harmony in the HM **do**
    evaluate training samples using Eq. (6.9) *% evaluate feature selection flag*
  end for
  //$CV_{Accuracy}$ and k_{opt}//
  for each $K \in n_1, \ldots, n_2$ **do**
    train the classifier using $(K - 1)$ number of fold *% train and test each fold of samples using feature selection flag*
    test the efficiency using the remaining fold
  end for
  until each of the K fold used once as a test fold
  for each $K \in n_1, \ldots, n_2$ **do**
    evaluate $CV\ training\ accuracy$ using Eq. (5.1) *% calculate CV training accuracy, CV test accuracy and value of optimum k*
    evaluate $CV\ test\ accuracy$ using Eq. (5.2)
    evaluate $SE(k)$ using Eq. (5.3)
  end for
  evaluate $CV_{accuracy}$ as:$max\{SE(k)\}$
  evaluate k_{opt}
  number of features in the candidate feature set
  //Fitness function//
  evaluate fitness function using Eq. (7.20) *% calculate fitness function using gene selection flag*
  //Group improvisation concept//
  Set the HS parameters as:
  $HMCR(T) = HMCR_{min} + \frac{(HMCR_{max} - HMCR_{min})}{iter_{max}} \times t$ *% calculate HMCR and PAR parameters*
  $PAR(T) = PAR_{min} + \frac{(PAR_{max} - PAR_{min})}{iter_{max}} \times t$
  number of groups as: $G(t) = \lfloor G_{max} - rand \times (G_{max} - G_{min}) \rfloor$ *% calculate the number of groups*
  while $g \le G(t)$ **do**
    select the member harmony for each group from HM in a random fashion *% calculate the number of harmony in each group*
    while (i $\le$ number of variables) **do**
      if $rand < HMCR(t)$ **then**
        if $rand < GMCR$ **then**
          choose a value from own group member harmony for variable i
          if $rand < PAR(t)$ **then**
            $z_i = z_i + bw \times ud(-1, +1)$, ud is uniformly distributed function
          end if

```
            else
              Choose a value from other groups for variable i
            end if
          else
            choose a random value
          end if
        end while
      end while
      accept the new harmony (solution) if better. % select the better harmony
    end while
    select the current best solution as optimum solution % select optimum solution
    END
```

$SI \leq 1$. A total number of 826961 LNCs have been produced this work. The multi-scale dot enhancement filtering has then been performed to remove the LNCs which are FPs. Three scales Gaussian with $\sigma_1 = 0.5$ mm, $\sigma_2 = 2$ mm, and $\sigma_3 = 8$ mm have been used to remove noise from the entire image. Thereafter, an LNC has been kept as PLNC if the particular LNC satisfies a predefined threshold criteria after multi-scale dot filtering stage. Here, a total number of 1135 nodule candidates have been retained as PLNCs using Algorithm 7.3 and the result could be seen in Fig. 7.4. The entire data set has then been randomly splitted into training and blind testing data set. The blind testing data set is needed for the independent testing to evaluate the performance of the proposed feature selection methodology.

In this work, K-fold cross-validation (with roughly equal sized folds) has been performed on the entire feature space. For each harmony in the proposed GrIHS-based feature selection technique, the model is trained K times, each time leaving one of the folds for testing. The trained model is tested finally on a blind testing data set. To judge the performance of the proposed technique, few important criteria have been considered: (a) CV accuracy, (b) size of the optimal feature set, (c) sensitivity, and (d) blind testing accuracy. High sensitivity, high blind testing accuracy and less number of features in the optimal set have been considered as an important condition to accomplish admirable performance.

Table 7.2 shows the result of ten independent runs of the proposed algorithm. The best set of selected features is shown in the shaded cell. Result shows that 97.59% sensitivity has been achieved with 97.78% blind testing accuracy using only 12 features. Finally, a region growing technique [50] has been applied to segment the nodule structure completely from the lung image. Figure 7.7 shows the examples of lung nodule segmentation. The white patch over the black background represents the segmented lung nodules. The original lung nodule and the segmented nodule patch are shown inside white circles. Figure 7.8 shows the segmented lung nodule with FPs where the nodule and FPs are shown inside white circles. The juxta-pleural nodules have also been considered in the proposed technique and the segmentation of juxta-pleural nodules are shown in Fig. 7.9. The proposed method has been compared with other applied methods on different LIDC data sets. Table 7.3 shows that the obtained result is very much comparable to other applied methods.

Table 7.2 Result of ten independent runs of the GrIHS-based feature selection and classification technique

Run	No. of features	Features	k	$CV_{Accuracy}$%	Sensitivity%	Blind test accuracy%
1	25	4,10,11,13,15,16,18,19,25,33,34,36,40,41,49,50,52,54,64,68,73,75,77,79,81	3	80.51	91.65	91.78
2	13	5,6,12,17,19,21,31,35,36,44,46,76,89	4	77.75	87.69	90.28
3	15	9,12,15,20,23,25,52,54,67,71,78,83,87,89,90	4	91.56	91.08	87.78
4	13	8,13,26,29,34,36,53,56,60,78,82,87,89	4	90.02	92.4	93.71
5	17	3,4,16,17,18,21,24,34,40,43,52,53,71,74,77,87,90	3	79.76	91.65	95.21
6	15	15,18,29,31,37,42,46,52,55,69,75,76,78,83,89	4	81.57	89.08	88.85
7	12	3,4,10,14,34,38,54,57,58,60,63,81	4	77.75	90.38	87.78
8	10	1,2,14,27,30,53,54,58,63,64	4	80.89	93.59	86.1
9	14	5,8,16,17,19,48,50,54,58,61,68,70,80,90	4	77.63	92.4	90
10	12	3,6,7,29,38,41,50,68,72,83,89,90	4	77.75	97.59	97.78
Mean	14	–	–	81.52	91.75	90.92

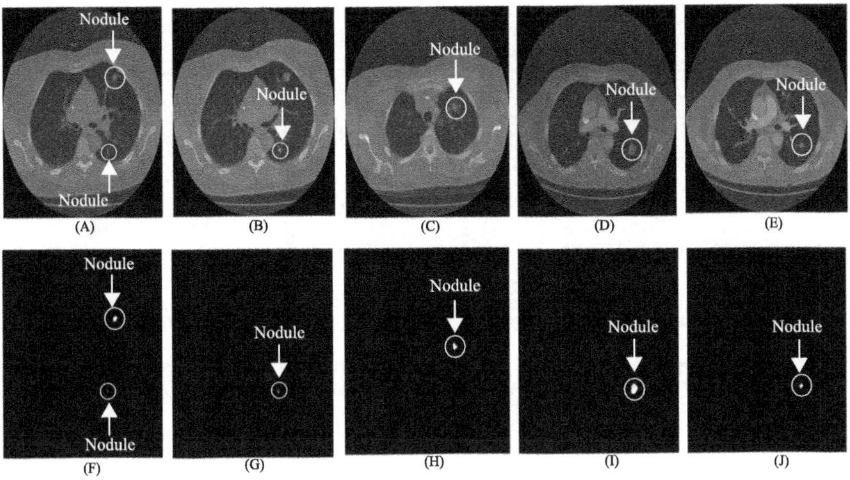

Fig. 7.7 Examples of segmented lung nodules

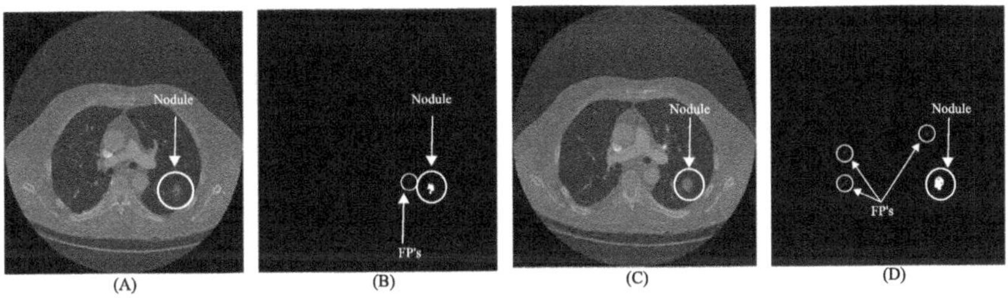

Fig. 7.8 Examples of segmented lung nodules with FPs

7.4.1 Result Analysis

The experimental results show that an optimal set of 12 features has been selected by the proposed method which is capable enough to classify the blind test samples with 97.59% sensitivity and 97.78% accuracy. Relation of the number of features and fitness with the progress of iterations is shown in Fig. 7.10. As fitness is obtained by weighted sum of CV error and number of features, it decreases with the progress of iterations. This trend of decreasing fitness function indicates that our proposed method could find a near-optimal subset of features with high blind test accuracy. However, the number of iterations to reach an optimal solution is problem dependent [56]. Therefore, 100 iterations have been chosen in this present case study. CV training accuracy, CV test accuracy and CV accuracy corresponding to the best result are also shown in Fig. 7.11. The optimal set of 12 features are shown in Table 7.4. The optimal set includes shape-based, color-based and texture-based features. Moreover, Table 7.2 suggests the possibility of finding more than one such set of important features. This demonstrates the presence of another such

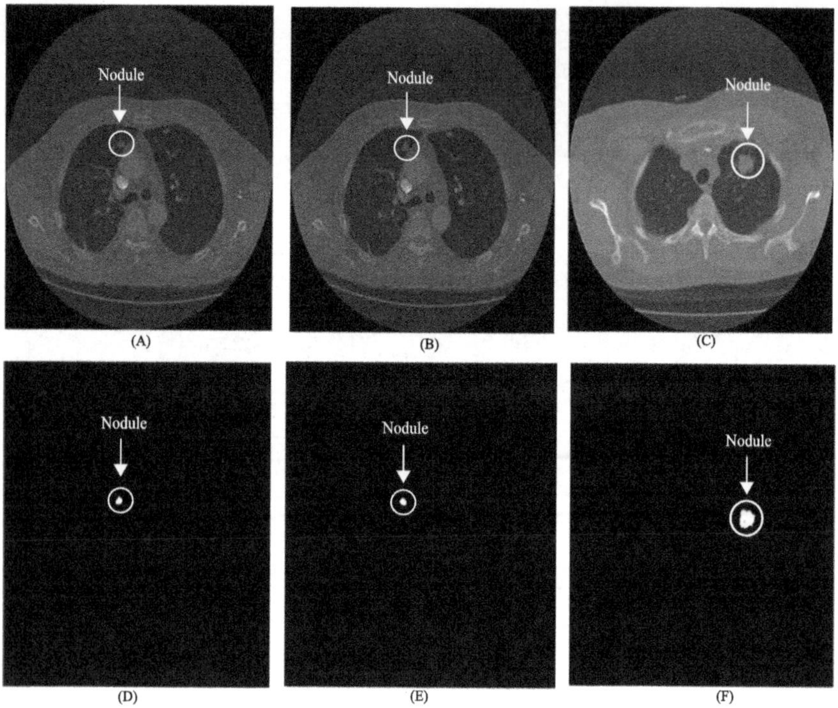

Fig. 7.9 Examples of juxta-pleural nodule segmentation

Table 7.3 Comparison of the proposed technique with other applied methods

Study	Method	Sensitivity%	Accuracy%
[38]	Shape-based detection	–	90.2
[51]	Dot enhancement filter	96.6	–
[52]	Texture features and SVM	96.31	96.54
[27]	Shape measurement	91.99	95.33
[53]	GLCM features + gain ratio	94.11	91.67
[54]	Improved random forest + bootstrap	92	90
[55]	Structural co-occurrence matrix + F-score	–	96.70
Proposed Technique	GrIHS + dynamic weight selection + A*k*-NN	97.59	97.78

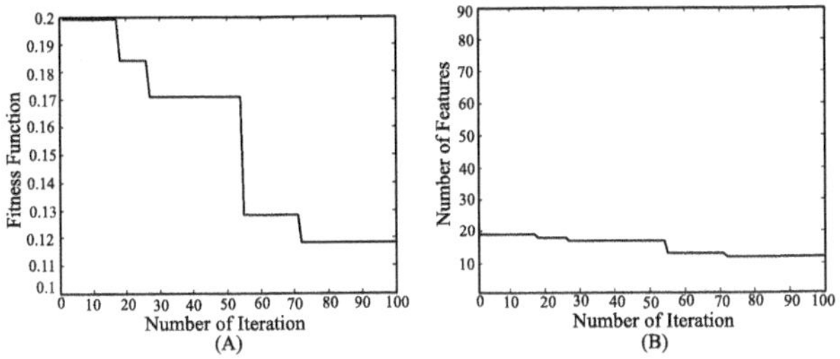

Fig. 7.10 Result of 100 iterations of GrIHS technique. (**a**) Fitness function (**b**) Number of features

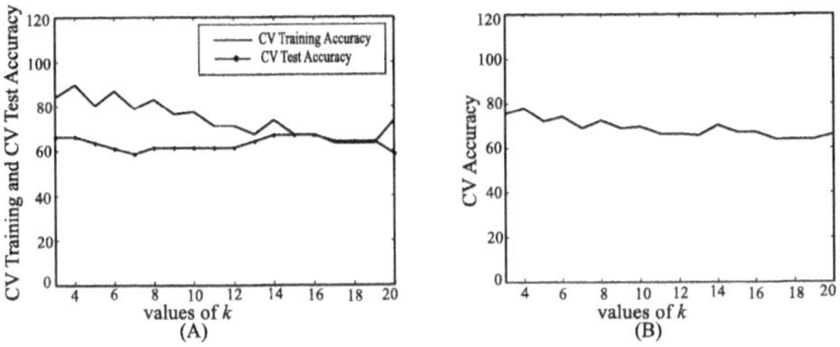

Fig. 7.11 Value of k (**a**) CV training accuracy and CV test accuracy vs. value of k (**b**) CV accuracy vs. value of k

Table 7.4 Optimum feature set

Feature index	Feature name	Feature nature
3	Mean intensity	Color-based
6	Max of shape index map	Shape-based
7	Mean of shape index map	Shape-based
29	Cluster prominence	Texture-based [$GLCM$ (0°)]
38	Sum variance	Texture-based [$GLCM$ (45°)]
41	Difference entropy	Texture-based [$GLCM$ (45°)]
50	Cluster shed	Texture-based [$GLCM$ (45°)]
68	Maximum probability	Texture-based [$GLCM$ (90°)]
72	Contrast	Texture-based [$GLCM$ (135°)]
83	Information measure correlation 2	Texture-based [$GLCM$ (135°)]
89	Cluster prominence	Texture-based [$GLCM$ (135°)]
90	Cluster shed	Texture-based [$GLCM$ (135°)]

feature sets which also can do the task well. The weights corresponding to the best result have been obtained as $\omega_1 = 0.8908$ and $\omega_2 = 0.1095$.

The proposed method has been compared with other reported feature-based classification methods applied to lung nodule classification. Table 7.3 portrays the fact that the results obtained from the proposed work are very much competitive to the results obtained from the other methods already reported in the literature and cited duly. The performance curve (ROC curve) of the proposed method and

Fig. 7.12 ROC curve

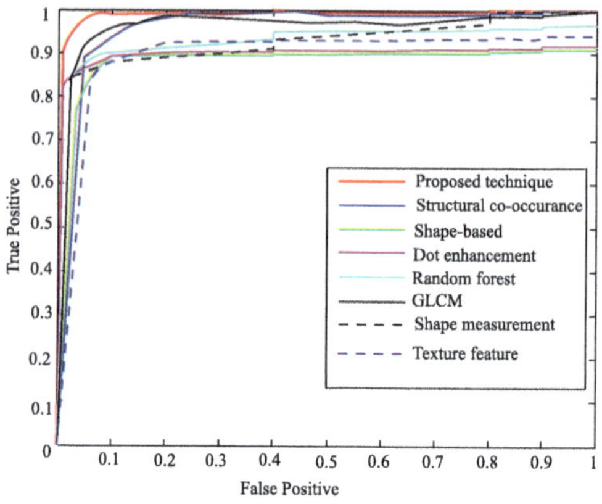

other relative methods (cited in Table 7.3) has been shown in Fig. 7.12 where it has been clearly observed that the ROC curve of the proposed method is not worse than others.

7.5 Summary

In this work, an adaptive weight selection strategy has been proposed for enhancing the lung nodule classification accuracy that actually requires an optimized feature subset. Here, the lung volume is first segmented using HU-based thresholding and a contour correction method. Then the nodule candidates have been detected using a volumetric shape index-based dot map calculation. The potential nodule candidates have then been extracted using multi-scale dot enhancement filtering technique. The features are then extracted from the detected potential nodule candidates. Finally, the extracted features have been used to develop GrIHS-based feature selection algorithm. This feature selection technique has been intertwined with adaptive *k*-NN classifier to grow simultaneously for selecting an optimal value of neighborhood for *k*-NN classifier. The proposed weight selection approach that integrates minimization of features for classification purpose is capable of selecting a near-optimal subset of useful features from a feature pool and thus increasing the blind test accuracy efficiently.

Moreover, the performance of the feature selection task can be improved by utilizing a large number of input features. The large number of features would help the group improvisation task to explore further and in turn would make the decisive feature selection process healthier. The proposed technique can then be applied to a larger data set for exploring a more robust feature selection and classification performance. The proposed technique can be used as a confirmatory option for the radiologists to interpret lung CT scans.

References

1. De Valk J, Eijkman E (1984) Analysis of eye fixations during the diagnostic interpretation of chest radiographs. Med Biol Eng Comput 22:353
2. Greenlee RT, Murray T, Bolden S, Wingo PA (2000) Cancer statistics, 2000. CA: Cancer J Clin 50:7
3. Kanazawa K, Kawata Y, Niki N, Satoh H, Ohmatsu H, Kakinuma R, Kaneko M, Moriyama N, Eguchi K (1998) Computer-aided diagnosis for pulmonary nodules based on helical CT images. Comput Med Imaging Graph 22:157
4. Armato SG, Giger ML, Moran CJ, Blackburn JT, Doi K, MacMahon H (1999) Computerized detection of pulmonary nodules on CT scans. Radiographics 19:1303
5. Hu S, Hoffman EA, Reinhardt JM (2001) Automatic lung segmentation for accurate quantitation of volumetric X-ray CT images. IEEE Trans Med Imaging 20:490
6. Li B, Christensen GE, Hoffman EA, McLennan G, Reinhardt JM (2003) Establishing a normative atlas of the human lung: intersubject warping and registration of volumetric CT images. Acad Radiol 10:255
7. Sluimer I, Prokop M, Van Ginneken B (2005) Toward automated segmentation of the pathological lung in CT. IEEE Trans Med Imaging 24:1025
8. De Nunzio G, Tommasi E, Agrusti A, Cataldo R, De Mitri I, Favetta M, Maglio S, Massafra A, Quarta M, Torsello M et al (2011) Automatic lung segmentation in CT images with accurate handling of the hilar region. J Digit Imaging 24:11
9. Li X, Wang X, Dai Y, Zhang P (2015) Supervised recursive segmentation of volumetric CT images for 3D reconstruction of lung and vessel tree. Comput Methods Programs Biomed 122:316
10. Lee Y, Hara T, Fujita H, Itoh S, Ishigaki T (2001) Automated detection of pulmonary nodules in helical CT images based on an improved template-matching technique. IEEE Trans Med Imaging 20:595
11. Ezoe T, Takizawa H, Yamamoto S, Shimizu A, Matsumoto T, Tateno Y, Iimura T, Matsumoto M (2002) Automatic detection method of lung cancers including ground-glass opacities from chest X-ray CT images. In: Medical imaging 2002: image processing, vol 4684. SPIE, pp 1672–1680
12. Li Q, Doi K (2004) New selective nodule enhancement filter and its application for significant improvement of nodule detection on computed tomography. In: Medical imaging 2004: image processing, vol 5370. SPIE, pp 1–9
13. Lin D-T, Yan C-R, Chen W-T (2005) Autonomous detection of pulmonary nodules on CT images with a neural network-based fuzzy system. Comput Med Imaging Graph 29:447
14. Paik DS, Beaulieu CF, Rubin GD, Acar B, Jeffrey RB, Yee J, Dey J, Napel S (2004) Surface normal overlap: a computer-aided detection algorithm with application to colonic polyps and lung nodules in helical CT. IEEE Trans Med Imaging 23:661
15. Pu J, Paik DS, Meng X, Roos J, Rubin GD (2010) Shape "break-and-repair" strategy and its application to automated medical image segmentation. IEEE Trans Vis Comput Graph 17:115
16. Hogeweg L, Sánchez CI, van Ginneken B (2013) Suppression of translucent elongated structures: applications in chest radiography. IEEE Trans Med Imaging 32:2099
17. Sun S, Guo Y, Guan Y, Ren H, Fan L, Kang Y (2014) Juxta-vascular nodule segmentation based on flow entropy and geodesic distance. IEEE J Biomed Health Inf 18:1355
18. Ciompi F, Jacobs C, Scholten ET, Wille MM, De Jong PA Prokop M van Ginneken B (2014) Bag-of-frequencies: a descriptor of pulmonary nodules in computed tomography images. IEEE Trans Med Imaging 34:962
19. Han H, Li L, Han F, Song B, Moore W, Liang Z (2014) Fast and adaptive detection of pulmonary nodules in thoracic CT images using a hierarchical vector quantization scheme. IEEE J Biomed Health Inf 19:648

20. Saien S, Moghaddam HA, Fathian M (2018) A unified methodology based on sparse field level sets and boosting algorithms for false positives reduction in lung nodules detection. Int J Comput Assist Radiol Surg 13:397
21. Gonçalves L Novo J, Campilho A (2016) Hessian based approaches for 3D lung nodule segmentation. Expert Syst Appl 61:1
22. Novo J, Gonçalves L Mendonça AM, Campilho A (2015) 3D lung nodule candidate detection in multiple scales. In: 2015 14th IAPR International Conference on Machine Vision Applications (MVA). IEEE, pp 61–64
23. Chen H, Zhang J, Xu Y, Chen B, Zhang K (2012) Performance comparison of artificial neural network and logistic regression model for differentiating lung nodules on CT scans. Expert Syst Appl 39:11503
24. Sun T, Wang J, Li X, Lv P, Liu F, Luo Y, Gao Q, Zhu H, Guo X (2013) Comparative evaluation of support vector machines for computer aided diagnosis of lung cancer in CT based on a multi-dimensional data set. Comput Methods Programs Biomed 111:519
25. da Silva Sousa JRF, Silva AC, de Paiva AC, Nunes RA (2010) Methodology for automatic detection of lung nodules in computerized tomography images. Comput Methods Programs Biomed 98:1
26. Setio AAA, Ciompi F, Litjens G, Gerke P, Jacobs C, Van Riel SJ, Wille MMW, Naqibullah M, Sánchez CI, Van Ginneken B (2016) Pulmonary nodule detection in CT images: false positive reduction using multi-view convolutional networks. IEEE Trans Med Imaging 35:1160
27. Filho AODC, Silva AC, de Paiva AC, Nunes RA, Gattass M (2017) 3D shape analysis to reduce false positives for lung nodule detection systems. Med Biol Eng Comput 55:1199
28. De Stefano C, Fontanella F, Marrocco C, Di Freca AS (2014) A GA-based feature selection approach with an application to handwritten character recognition. Pattern Recogn Lett 35:130
29. Krisshna NA, Deepak VK, Manikantan K, Ramachandran S (2014) Face recognition using transform domain feature extraction and PSO-based feature selection. Appl Soft Comput 22:141
30. Inbarani HH, Azar AT, Jothi G (2014) Supervised hybrid feature selection based on PSO and rough sets for medical diagnosis. Comput Methods Programs Biomed 113:175
31. Ramos CC, Souza AN, Chiachia G, Falcão AX Papa JP (2011) A novel algorithm for feature selection using harmony search and its application for non-technical losses detection. Comput Electric Eng 37:886
32. Huang Y-F, Lin S-M, Wu H-Y, Li Y-S (2014) Music genre classification based on local feature selection using a self-adaptive harmony search algorithm. Data Knowl Eng 92:60
33. Shreem SS, Abdullah S, Nazri MZA (2016) Hybrid feature selection algorithm using symmetrical uncertainty and a harmony search algorithm. Int J Syst Sci 47:1312
34. Inbarani HH, Bagyamathi M, Azar AT (2015) A novel hybrid feature selection method based on rough set and improved harmony search. Neural Comput Appl 26:1859
35. Kong X, Gao L, Ouyang H, Li S (2015) A simplified binary harmony search algorithm for large scale 0–1 knapsack problems. Expert Syst Appl 42:5337
36. Kar S, Das Sharma K, Maitra M (2020) Adaptive weighted aggregation in Group Improvised Harmony Search for lung nodule classification. J Exp Theor Artif Intell 32:219
37. Youssif AA, Hussein SA, Ibrahim AS (2011) Three-dimensional detection of pulmonary nodules in chest ct images. Comput Inf Sci 4:2
38. Ye X, Lin X, Dehmeshki J, Slabaugh G, Beddoe G (2009) Shape-based computer-aided detection of lung nodules in thoracic CT images. IEEE Trans Biomed Eng 56:1810
39. Worthington PL, Hancock ER (2001) Surface topography using shape-from-shading. Pattern Recogn 34:823
40. Li Q (2009) Selective enhancement filters for nodules, vessels, and airway walls in two-and three-dimensional CT scans. Med Phys 30:1810
41. Koenderink JJ Van Doorn AJ (1992) Surface shape and curvature scales. Image Vision Comput 10:557
42. Yoshida H, Nappi J (2001) Three-dimensional computer-aided diagnosis scheme for detection of colonic polyps. IEEE Trans Med Imaging 20:1261

43. Geem ZW, Kim JH, Loganathan GV (2001) A new heuristic optimization algorithm: harmony search. Simulation 76:60
44. Geem ZW (2008) Novel derivative of harmony search algorithm for discrete design variables. Appl Math Comput 199:223
45. Mahdavi M, Fesanghary M, Damangir E (2007) An improved harmony search algorithm for solving optimization problems. Appl Math Comput 188:1567
46. Sharma KD, Chatterjee A, Rakshit A (2010) Design of a hybrid stable adaptive fuzzy controller employing Lyapunov theory and harmony search algorithm. IEEE Trans Control Syst Technol 18:1440
47. Sharma KD (2013) Stable fuzzy controller design employing group improvisation based harmony search algorithm. Int J Control Autom Syst 11:1046
48. Boroczky L, Zhao L, Lee KP (2006) Feature subset selection for improving the performance of false positive reduction in lung nodule CAD. IEEE Trans Inf Technol Biomed 10:504
49. Armato III SG McLennan G, Bidaut L, McNitt-Gray MF, Meyer CR, Reeves AP, Zhao B, Aberle DR, Henschke CI, Hoffman EA et al (2011) The lung image database consortium (LIDC) and image database resource initiative (IDRI): a completed reference database of lung nodules on CT scans. Med Phys 38:915
50. Gonzalez RC (2009) Digital image processing. Pearson Education India
51. Lee SLA, Kouzani AZ, Hu EJ (2010) Random forest based lung nodule classification aided by clustering. Comput Med Imaging Graph 34:535
52. Akram S, Javed MY, Hussain A, Riaz F, Usman Akram M (2015) Intensity-based statistical features for classification of lungs CT scan nodules using artificial intelligence techniques. J Exp Theor Artif Intell 27:737
53. Sebatubun M, Haryawan C, Windarta B (2018) Classification of ground glass opacity lesion characteristic based on texture feature using lung CT image. J Exp Theor Artif Intell 30:203
54. Li X-X, Li B, Tian L-F, Zhang L (2018) Automatic benign and malignant classification of pulmonary nodules in thoracic computed tomography based on RF algorithm. IET Image Process 12:1253
55. Rodrigues MB, Da Nobrega RVM, Alves SSA, Reboucas Filho PP, Duarte JBF, Sangaiah AK, De Albuquerque VHC (2018) Health of things algorithms for malignancy level classification of lung nodules. IEEE Access 6:18592
56. Engelbrecht AP (2006) Fundamentals of computational swarm intelligence. John Wiley & Sons

Chapter 8
Conclusion

Abstract The intelligent approaches for the detection and classification of different carcinogenic diseases are presented in this book, with an emphasis on information at the gene, cell, and tissue levels. The suggested frameworks achieve strong outcomes in microscopic image investigations and DNA microarray data processing by employing intelligent assessment criteria. Important contributions include the study of gene subset identification, improved gene selection using an adaptive k-NN strategy based on PSO, and sophisticated methods for lung nodule and leukocyte detection and categorization. These methods offer medical professionals a helpful second opinion and perform on par with or better than existing methods. The book also highlights future research possibilities, stressing the value of standardized diagnostic methods and the promise of deep learning networks in biomedical image analysis. It also emphasizes the applicability of microarray technology in clinical microbiology and multimodal medical imaging in advancing cancer research and intelligent diagnostics.

In this book, some intelligent methodologies for detection and classification of different types of carcinogenic diseases have been proposed and discussed in detail. The detection and classification methodologies have been constructed on a strong foundation of gene level information, cell level information, and tissue level information that permits to exploit different intelligent evaluation criteria in different search stages. The goal of the book is to perform different intelligent diagnosis techniques with the dual objective to achieve good classification performance and to discover an optimal subset of features (genes in some cases) relevant for the application domain.

To evaluate the proposed framework, extensive analysis has been conducted on three DNA microarray data sets (SRBCT, ALL_AML, and MLL). The proposed approach obtains comparable and better results than other methods presented in the literature. The book also shows that the proposed approach is robust and effective in finding optimal subsets of informative genes. The framework has also been validated with some detail studies in the domain of microscopic blood smear image and

K. Das Sharma et al., *Intelligent Computing in Carcinogenic Disease Detection*, Computational Intelligence Methods and Applications,
https://doi.org/10.1007/978-981-97-2424-6_8

CT image analysis. The proposed approaches achieve comparable performances on detection and classification accuracies with other methods reported in the literature.

In this context, the contributions of this book are reiterated as follows:

- The performance of finding an informative gene subset from high-dimensional DNA microarray data sets has been investigated by employing filter and wrapper approaches.
- The performance of gene selection capability has been improved by employing a PSO-based adaptive k-NN methodology. The uniform characteristic and usefulness of the selected informative genes have been reconfirmed by an SVM classifier with different kernels.
- The intelligent detection and classification of leukocytes have been performed by using a WATPSO technique. The weights assigned with dual objectives have been proposed to transpose experimentally for obtaining the optimal solution.
- An adaptive weight selection strategy has been embedded in GrIHS-based feature selection technique for the detection and classification of lung nodules. To enhance the classification accuracy, the adaptive k-NN strategy has also been employed in the proposed algorithm.

Therefore, in essence, some intelligent cancer detection and classification methodologies have been proposed for different types of medical data sets in this book. The proposed methodologies can act as a second opinion for radiologists, hematologists, and doctors for the accurate detection of carcinogenic diseases.

8.1 Future Research Directions

The future research area of the application of microarray technology is clinical microbiology. The low- or middle-density microarrays can help in the simultaneous assessment of large numbers of microbial genetic objects. The future research in this domain should guide the possibility of developing simplified methods for an easy diagnosis of the cancer subgroups. The possibility in future would be to design specialized bio-chips for these very purposes.

Furthermore, to realize the advantages of intelligent methodologies in cancer diagnosis and classification, the improvement of the ancillary technologies should be developed. New intelligent methodological platforms should come up in tandem with the statistics and software for analysis and data mining technologies in early cancer detection. Then only more intelligent technical and analytical procedures will benefit the patients in a large context.

In addition to that standardized diagnostic techniques employing cell level and tissue level information should be formulated in cooperation with different laboratories working for generating meta-profiles in this direction of cancer research. In this context, deep learning network should come up to detect pathology in biomedical images. Deep learning will be useful in improving accuracy and increasing efficiency.

The future research direction should also be attributed to deep convolution neural network. The multi-stream architecture of deep convolution neural network can accommodate multiple sources of information of the input that will help greatly in medical image segmentation. In future, deep convolution neural networks can also be used to leverage the intrinsic structure of medical images.

The future research should also be directed in analyzing multimodal medical imaging. This approach can provide a large amount of information from gene level, cell level, and tissue level medical data. Multimodal imaging techniques should play an important role in clinical applications. The future research in multimodal medical imaging, mainly in the field of oncology, should impact the way medicine is performed today. Future research should demonstrate the efficient uses of the imaging information provided by the modalities to affect the future of intelligent diagnostic.

Index

K. Das Sharma et al., *Intelligent Computing in Carcinogenic Disease Detection*, Computational Intelligence Methods and Applications,
https://doi.org/10.1007/978-981-97-2424-6

GPSR Compliance

The European Union's (EU) General Product Safety Regulation (GPSR) is a set of rules that requires consumer products to be safe and our obligations to ensure this.

If you have any concerns about our products, you can contact us on ProductSafety@springernature.com

In case Publisher is established outside the EU, the EU authorized representative is:

Springer Nature Customer Service Center GmbH
Europaplatz 3
69115 Heidelberg, Germany

Batch number: 10725971

Printed by Printforce, the Netherlands